Plasma Applications for Material Modification

Plasma Applications for Material Modification

From Microelectronics to Biological Materials

edited by

Francisco L. Tabarés

JENNY STANFORD
PUBLISHING

Published by

Jenny Stanford Publishing Pte. Ltd.
Level 34, Centennial Tower
3 Temasek Avenue
Singapore 039190

Email: editorial@jennystanford.com
Web: www.jennystanford.com

British Library Cataloguing-in-Publication Data
A catalogue record for this book is available from the British Library.

Plasma Applications for Material Modification:
From Microelectronics to Biological Materials

ISBN 978-981-4877-35-0 (Hardcover)
ISBN 978-1-003-11920-3 (eBook)

Contents

Preface

Plasmas, also called the fourth state of matter, are ubiquitous in modern societies. You will find them from your TV screen to the headlights of your car or street lighting, just to mention a few examples. They are also present in nature, as in lightning and flames, but what makes plasmas paramount is that they represent the predominant state of matter in the known universe, as it may account for up to 99% of its material content.

This book addresses just one of the many applications that plasmas have found in our society, surface modification.

Perhaps some of the readers of this book are not fully aware of the crucial role that surface properties play in day-to-day life. Acting in a rather independent way that the bulk of the material does, surfaces represent a boundary between two media, one of them being typically the atmosphere. Characteristics such as reflectance, friction, corrosion and passivation, electron emission, catalytic effects, and many others are genuine surface properties affecting the performance of many devices of use in today's life. Furthermore, the addition of coatings on common materials allows for the tailoring of specific application-driven attributes that the original material did not have. Perhaps the coating of iron frying pans with Teflon to prevent the sticking of the food while cooking may serve as a well-known example of the use of target-driven coatings by plasma techniques.

More recently, plasmas have met life. There is a plethora of new applications of plasmas to biological targets, including medicine, agriculture, food processing and the like that, from their staggering initial attempts, are proliferating quickly and relentlessly. Again, in this emerging field, the interaction of the plasma with the biological target is restricted to the surface.

In this context, the purpose of this book is to give a unified and comprehensive presentation of the plasma-based techniques used for surface modification and their implementation in practical applications. Rather than dwelling on very specific issues of each application, the goal of the book is to provide a solid background on plasma physics and chemistry and to relate them to the resulting changes in the observed surface properties. After a general introduction in Chapter 1, the book is divided into seven chapters that describe the different kinds of plasmas and their fields of application, from fusion reactors to dentistry. Each chapter is self-contained and includes both fundamentals and the latest research results. The book should therefore prove useful for undergraduate and graduate students and researchers working on plasmas and surface physics and engineering .

This book would not have happened if it were not for the extraordinary work of all contributing authors, the thorough revision from a selected group of reviewers, and the continuous support from the whole team at Jenny Stanford.

Enjoy it!

Francisco L. Tabarés

Madrid, 2021

Chapter 1

Introduction: Cold Plasmas and Surface Processing

F. J. Gordillo[a] **and F. L. Tabarés**[b]
[a]*Instituto de Astrofísica de Andalucía (IAA – CSIC), Glorieta de la Astronomía s/n, 18008 Granada, Spain*
[b]*Laboratorio Nacional de Fusion, CIEMAT, Av Complutense 40, 28040 Madrid, Spain*
tabares@ciemat.es

The first time the word "plasma" appeared in print in a scientific text related to the study of electrical discharges in gases dates back to 1928. That year Irving Langmuir published his article "Oscillations in Ionized Gases" in *Proceedings of the National Academy of Sciences of the States United*. It was a baptism of the predominant state of matter in the known universe (it is estimated that up to 99% of matter is plasma), although not on our planet, where the conditions of pressure and temperature make normal the states of matter—solid, liquid, and gas—that in global terms are exotic. It is enough to add energy to the solid (in the form of heat or electromagnetic radiation) for it to turn into a liquid state, from which gas is obtained through an additional supply of energy. If we continue adding energy to the gas, we will partially or totally ionize it, that is, we will remove electrons from the atoms or molecules that constitute

Plasma Applications for Material Modification: From Microelectronics to Biological Materials
Edited by Francisco L. Tabarés

ISBN 978-981-4877-35-0 (Hardcover), 978-1-003-11920-3 (eBook)
www.jennystanford.com

it. In this way, we reach a new state of matter, plasma, made up of free electrons, atoms and molecules (electrically neutral particles), and ions (endowed with a positive or negative electric charge). The energy needed to generate a plasma can be supplied in several ways: through heat from a combustion process; through the interaction between laser radiation and a solid, a liquid, or a gas; or through electrical discharges in gases, in which free electrons take energy from the applied electric field and lose it through excitation and ionization processes of the atoms and molecules in the gas.

The light emitted by a plasma, its characteristic emission spectrum, is determined by the type of atoms, molecules, and ions that form it. These components, when de-energized, emit electromagnetic radiation, visible or not. One of the peculiarities of plasmas is that they conduct electricity. On a macroscopic scale, plasmas are, however, electrically neutral, since the number of positive and negative charges is similar. Thus, the flame produced by burning candle wax in combination with oxygen from the air—a typical example of a plasma that is very little ionized—can conduct electricity. A graphical overview of the different kinds of plasmas according to their microscopic parameters (electron density and temperature) is displayed in Fig. 1.1.

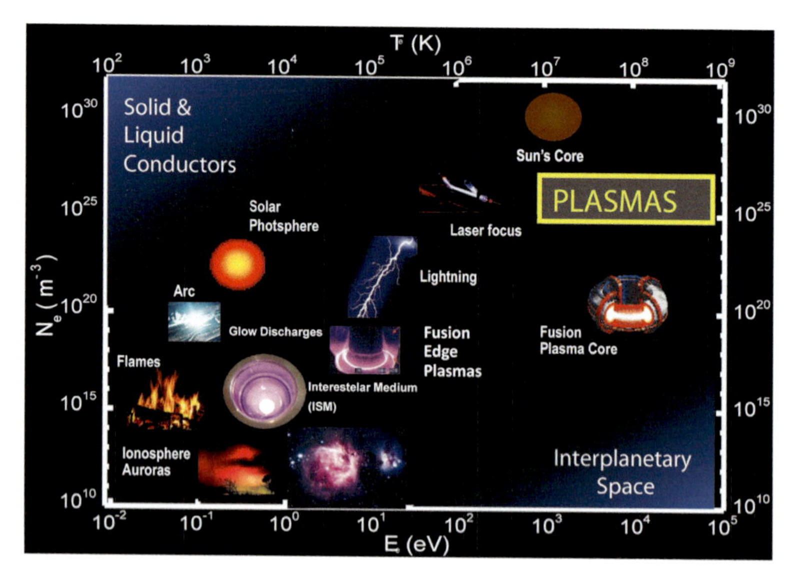

Figure 1.1 Overview of plasmas and the wide range of their microscopic parameters (electron energy and electron density).

The study of common natural phenomena in our world has taught us that lightning (Fig. 1.2), auroras, the ionosphere, and the recently discovered electrical discharges in the upper atmosphere (between 40 and 90 km high) are different types of natural plasmas present in the gaseous envelope of our planet. Beyond the Earth, there are plasmas in the Sun and other stars, in the solar wind, in the tails of comets, and in the interstellar space. The first observations related to plasmas go back to the experiments of Georg Christoph Lichtenberg, a professor at the University of Göttingen in the last third of the eighteenth century and today most remembered as a writer. By placing an insulating material between a pointed electrode and a metal plate and subjecting it to high electrostatic stress, he observed beautiful radiant patterns with tree-like shapes. These patterns were due to the dielectric breakdown of the material. The first attempts to explain Lichtenberg's observations were made by Michael Faraday, who dedicated some years of his life (1816–1819) to the study of the characteristics of matter when its temperature increases, although he did not elucidate the possible existence of a new state of matter beyond the gaseous state. Another English physicist, William Crookes, discovered in 1879 a green "radiant matter" with striated patterns that appeared when applying voltages between electrodes installed inside a glass tube, filled only with the air that remained after it was emptied. In addition, near the cathode, he observed a dark region, the so-called Crookes' dark zone. These observations led him to postulate the existence of a fourth state of matter. He conjectured that it was made up of gas molecules endowed with an electrical charge, that is, ions. Before these works, in 1857, Werner von Siemens had already patented an industrial process that used plasmas for the production of ozone: oxygen flowed through an annular electrical discharge between two concentric electrodes, one of which had an insulator material attached as a dielectric barrier. Although Siemens ignored the ultimate scientific reasons on which the method was based, it was very efficient and cost effective.

J. J. Thomson's work on cathode rays in electrical discharges in gases and his discovery of the electron in 1897 earned him the 1906 Nobel Prize in Physics. His was a remarkable contribution to the knowledge of the structure of atoms (composed of a positive nucleus and surrounding negatively charged electrons), and in doing so, he helped to clarify the nature of plasmas. The first attempt to give an overview of the physics of gas discharges was made by Johannes Stark in his book *Elektrizität in Gasen*, published in Germany in 1902.

Figure 1.2 Lightning, together with flames, is the most conspicuous kind of plasma in nature.

1.1 Types of Plasmas

Classifying the diversity of types of plasmas that exist in nature or that can be generated artificially is not easy, since it is risky to choose isolated parameters that serve as criteria to establish the differences. Despite these difficulties, we can venture into a first classification of the types of plasmas, one that takes into account their thermal equilibrium, that is, whether or not the temperature or the average energy of the particles that make it up is the same for each type of particle.

All particles have the same temperature (thermal equilibrium) for stellar interior plasma or for its terrestrial analogs, deuterium-tritium fusion plasmas and the impurities generated in experimental controlled nuclear fusion devices, like JET[a] and ITER[b] (see Fig. 1.4). The plasma inside the stellar interior is usually made up of a high proportion of ionized particles: the number of electrons, and of ions, is similar to that of neutral particles. These plasmas are also called hot or thermal plasmas, since the temperature inside them reaches millions of degrees (10^7°C–10^9°C), the same for electrons as for heavy species.

There are other types of thermal plasmas, with certain industrial applications, that are generated at high pressures, above 133 mbar, just over a tenth of an atmosphere, although their temperatures (10^4°C–10^5°C) are much lower than those of fusion plasmas. Plasma torches for surface treatment or plasma lamps that produce high-intensity discharges for street lighting or headlights of high-end cars are such plasmas.

When the gas pressure is low or the electrical voltage applied in the discharge is high, the electrons in the plasma acquire, in the time between collisions with other plasma particles, kinetic energies higher than the energy associated with the random thermal movement of the neutral particles (atoms and molecules) of the plasma. We can then attribute some degree of thermal equilibrium deviation to plasmas, since electrons, ions, and neutral particles have different "temperatures" or average kinetic energies. Please note that it only makes sense to talk about temperature when the energy distribution of the particles in question is limited to a certain statistical function, the Maxwellian one. This is not usually the case in plasmas produced at a low pressure and with a small degree of ionization between 10^{-6} and 10^{-4}.

Nonthermal plasmas, also known as cold plasmas, are characterized by the fact that the temperature of heavy species (neutral particles and ions) is close to room temperature (25°C–100°C). Instead, the electronic temperature is much higher

[a]The Joint European Torus.

[b]Originally, International Thermonuclear Experimental Reactor.

(between 5000°C and 10^5°C). Cold plasmas usually occur at a low pressure ($p < 133$ mbar) in reactors with very different geometries. Such reactors generate plasmas through direct current, radio frequency, microwave, or pulsed discharge systems.

There are special types of cold plasmas, produced in so-called corona and dielectric barrier discharges, that are generated at atmospheric pressure by using pulses between 10^{-6} s and 10^{-9} s. In these types of discharges, highly energetic electrons are produced that, due to the shortness of the pulses used, have little time to exchange energy with their surroundings. This establishes a strong temperature gradient between the electrons and the heavy species in the plasma.

The values of density and electronic temperature, two of the main parameters that characterize plasmas, cover a wide spectrum (see Fig. 1.1). Thus, the electron density varies between 1 electron/cm^3 and 10^{25} electrons/cm^3; that is, it even exceeds the concentration of electrons in metals. On the other hand, the average free path of the particles in a plasma, that is, the average distance covered before a particle collides with another particle in the plasma, can range from tens of millions of kilometers to just a few microns.

1.2 Cold Plasma in the Industry

Cold plasmas are very useful for many technical applications because, since they are not in thermal equilibrium, it is possible to control the temperature of ionic and neutral species on the one hand and of electrons on the other. However, the high energy of the electrons constitutes the genuine determining factor when initiating many chemical reactions that in thermally activated media would be very inefficient, if not impossible.

Industrial applications of cold plasmas make up a very important part of the productive infrastructure of advanced countries. In cold plasmas, a large number of and diverse highly energetic reactive species are generated that activate physical and chemical processes that are difficult to achieve in ordinary chemical environments. These species include photons in the visible and ultraviolet range, charged particles (electrons and ions), highly reactive neutral species (such as free radicals or oxygen), fluorine and chlorine atoms, excited atomic and molecular species, and excimers and monomers (an excimer is an electronically excited molecule that lacks a stable ground state; a monomer is a highly reactive chemical subunit that can bind to other equals to form polymers).

Thanks to cold plasmas, certain industrial processes are carried out more efficiently and cheaply, thereby reducing pollution and the toxic waste generated. The advantages of the industrial use of cold plasmas are perfectly illustrated in the comparison that W. Rakowski published in 1989 between the resources needed to dye cotton fabrics with a current chemical method that uses chlorine and those required for an equally effective procedure that uses cold plasmas at a low pressure (2.5–7 mbar). Modifying 20 tons/year of wool using the second method saved 27,000 m^3 of water, 44 tons of sodium hypochlorite, 16 tons of sodium bisulfite, 11 tons of sulfuric acid, and 685 MW of electrical power. Furthermore, the ordinary chemical process produced toxic residues causing different diseases in the workers. Comparing the energy costs of producing 1 kg of dyeable wool fabric gave figures of 7 kW/kg for the traditional chlorination process versus only 0.3 or 0.6 kW/kg when cold plasma treatment, produced at a low pressure, is used.

1.3 Cold Plasma Chemistry

The chemistry of cold plasmas, or cold chemistry, so called because of the low temperature (generally less than 100°C) of heavy plasma species, can be of the homogeneous type or the heterogeneous type. It will be of a homogeneous type when the reactions take place in the gas phase, as in the synthesis of ozone or in the elimination of sulfides and nitrides present in waste gases. It will be of a heterogeneous type when the plasma interacts with a solid or liquid surface.

In the plasma–solid surface interaction processes, three categories are recognized: erosion, deposition, and physicochemical alteration. Erosion is understood as plasma-assisted wear of a surface by simultaneous sputtering. It is of great interest to the microelectronics industry because it erodes the material anisotropically, that is, the material ends up having a different width and height, while typical of ordinary chemical techniques is isotropic carving. In plasma-assisted vapor phase chemical deposition processes the material is added to the surface in the form of a thin layer. Finally, solid surfaces treated with plasmas undergo physicochemical changes as an effect of the radiation processes and of the particles coming from the plasma that act on it.

The scientific and technical interest in plasma-surface interaction processes arose from a work by Jerome Goodman, published in 1960. He argued that a sheet of material deposited from a plasma could be useful and not just an annoying residue. Specifically, Goodman observed that the 1 μm thick deposit of plasma-polymerized styrene exhibited valuable dielectric properties. From that moment on, the synthesis of polymeric materials under the influence of cold plasmas, or plasma polymerization, ceased to be an undesirable by-product, already observed in 1874 by de Wilde and Thenard, to become one of the plasma material treatments with a greater number and diversity of applications.

Plasmas produced in corona discharges at atmospheric pressure have been used for the surface treatment of materials, although, because they are very inhomogeneous, they have been gradually replaced by plasmas generated in low-pressure luminescent discharges (between 0.013 mbar and 13 mbar). However, in the last 15–20 years, barrier and luminescent discharges have also been used that generate plasmas at atmospheric pressure in a homogeneous regime, which makes them practical and economically competitive compared to those produced at a low pressure in many processes that use cold plasmas to treat large surfaces.

Instead of immersing the sample in the plasma reactor, the active species responsible for the enhanced chemistry can be generated in an external plasma and then applied to the surface of interest. This technique is thoroughly described in Chapter 7.

1.4 Microelectronics

Since the mid-1960s, the intense demand by the microelectronic industry of circuits with increasing scales of integration has been the main stimulus for the development of treatment methods of plasma and thin layered materials. Thus, plasma-assisted dry eroding, which enabled the massive creation of anisotropic patterns on silicon wafers, evolved in the 1970s. Furthermore, during that same decade, it became common for certain plasma diagnosis methods to be used in the production lines of integrated circuits, among them emission light spectroscopy, which, with the ultimate intention of controlling and optimizing manufacturing, allowed the deepening of the knowledge of the kinetics of plasma and of the mechanisms of erosion and deposition. Plasma diagnosis techniques were used in conjunction with the new tools that appeared at that time for the analysis of surfaces, such as X-ray photoelectron spectroscopy and Auger electron spectroscopy.

In early 2006, Intel announced the commercialization of a new microprocessor based on an integration technique that has achieved dimensions as small as 45 nm in circuit elements. Such a degree of miniaturization would not have been possible without a substantial investment in research dedicated to optimizing the three types of cold plasma–based processes mentioned above. Today, up to two-thirds of the stages involved in the semiconductor manufacturing process are based on the use of cold plasmas. Far from reaching a stable situation, the trend is growing steadily and has even extended to related industrial sectors, for example, the production of photovoltaic cells by the synthesis of thin sheets of amorphous or microcrystalline silicon with thin-sheet plasma-assisted deposition techniques. On the other hand, the use of microplasmas, such as those produced in hollow cathode microdischarges, allows millions of microholes (about 50 mμ in diameter) to be made in integrated circuit boards in 1 hour; Thus, the density of components in the circuits of microelectromechanical devices is significantly increased.

Closely related to microelectronics, but not restricted to it, growing thin films with tailored properties is a subject on which cold plasmas have proven outstanding. Two kinds of plasma-based techniques are routinely used: plasma-enhanced chemical vapor deposition and physical vapor deposition. These techniques are thoroughly reviewed in Chapters 2 and 3, respectively.

1.5 Surface Treatments with Cold Plasmas

The energy and electric charge of the particles that make up a plasma change abruptly when they come into contact with solid surfaces. Energy is transferred to surfaces. On the one hand, plasmas can activate a surface, that is, make its molecules bond with the molecular components of other substances; on the other, by choosing the gases with which the plasmas are generated properly, they will serve to cover the surface with a certain material.

If cold plasmas can be generated inside large vacuum vessels, such as in particle accelerators and fusion devices, their application leads to a change in the surface composition, which is beneficial for the control of outgassing, secondary electron emission, and impurity generation, among others. Figure 1.3 shows one example of this in a fusion device. By inserting an electrode and selecting the gas and discharge parameters, the conditioning of the first wall is realized in preparation of the production of hot plasmas. A review of this topic is given in Chapter 6.

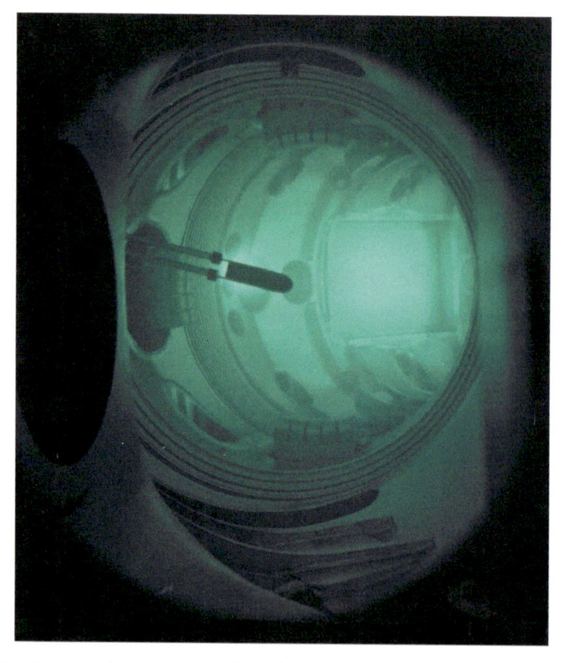

Figure 1.3 A cold plasma is produced in the chamber of a fusion device for wall conditioning through physical and chemical processes.

In general, a surface is treated with a cold plasma to activate it. An example of surface activation by plasma is the humidification of cotton fibers, which improves the adhesive capacity of the dye molecules, which makes it possible to have fabrics with higher-quality colors. In addition, the staining itself develops more quickly than through ordinary chemical methods. The ability of cold plasmas to modify surfaces is due to several factors. One of them is the high average energy of the free electrons present in the plasma (1–10 eV), enough to break chemical bonds. Furthermore, the unique characteristics of surface treatments using cold plasmas derive from the effects produced on the surface by photons and active species from plasma, which penetrate to a depth a few hundred angstroms ($\sim 10^{-8}$ m) to about 10 μm. Due to this, the properties of the interior of the material remain unchanged. The treatment of surfaces with cold plasmas allows functional activation, or "functionalization," of the surfaces, something that chemical treatments cannot offer. An example: the treatment of synthetic surfaces with a plasma of pure oxygen generates hydroxyl, carbonyl, and ester groups in it. Or, put in another way, it promotes the formation of chemically functional molecular groups that improve interfacial adhesion and, thus, the surface treated with an oxygen plasma will be receptive to subsequent treatments. These processes today have a significant economic impact in many sectors. The functionalization of technical fabrics to make them hydrophobic and impervious to moisture or oil is carried out by means of cold plasmas generated at atmospheric pressure in dielectric barrier discharges. The cold energy increase of the surface energy of polymeric materials improves the adhesion of certain coatings, for example, a sheet of very thin insulating material, about 40 nm thick, which acts as a barrier against penetration. Chapter 4 offers a comprehensive review of the generation of atomic species by RF plasmas for surface chemical tailoring.

1.6 Controlled-Fusion Plasmas

The generation of energy from the fusion of light atoms in a controlled way is a longstanding dream dating back to the times following World War II and H-bomb realization. Although several approaches to the practical implementation of such kinds of energy exist, the use of magnetically confined plasmas of deuterium and tritium—the heaviest isotopes of hydrogen—stands out as the most advanced concept in terms of maturity and international engagement. The present construction of the ITER tokamak (Fig. 1.4) in the south of France under the financial, technical, and scientific support of the main world's countries (encompassing more than the 60% of the human population) is just an example of the actual commitment toward this new, highly promising kind of energy.

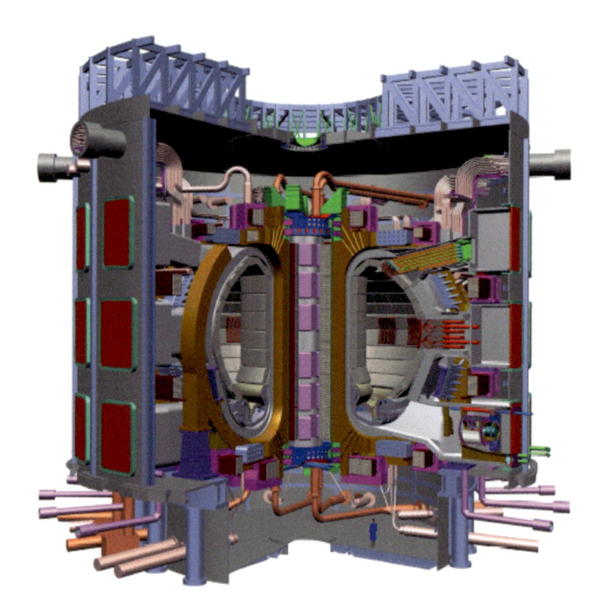

Figure 1.4 ITER will be the largest experimental fusion device in the world on route to commercial fusion reactors.

Chapter 5 offers a rather comprehensive introduction to fusion plasmas, so only a few, very general considerations are addressed here. At the temperatures required to fuse D and T ions after overcoming the coulomb electrostatic repulsion ($\sim 10^8$ C) a fully ionized plasma is produced. Under ignition, this plasma needs to be confined long enough in order to deliver the amount of fusion energy required for the reaction to self-maintain, the so-called Lawson tripe product criterion:

$$n\tau T \geq 10^{21}\ \text{m}^{-3}\text{s KeV},$$

where n stands for the density of particles, T their temperature, and τ the characteristic confinement time of the energy in the plasma.

In a DT plasma, energy is released in the form of neutrons (80%) and alpha particles (20%). Only the latter are confined to the magnetic field, thus colliding with the D and T ions and heating them back to the required temperature for self-sustainment of the process. However, in a steady-state situation, all this He^{2+} ash and its associated energy has to be extracted from the burning plasma through material elements, and hence, plasma wall interaction becomes an issue.

Since the region of the plasma close to material surfaces is significantly cooler than the core (mostly due to the strong radiation of impurity species), the physical and chemical processes taking place at the exposed surfaces are not too different from those involved in cold plasma material processing. However, the presence of a confining magnetic field fosters the transport of the released (eroded) material along the device, leading to material mixing issues, as will be described in Chapter 5

1.7 Medical and Biomedical Applications

In 1969, John R. Hollahan's group experimentally demonstrated that with cold ammonia plasmas or mixtures of nitrogen and hydrogen, amino groups ($-NH_2$) were produced, which, by adhering to the surface of different types of polymers, created materials compatible with blood. Since then, the use of cold plasmas to optimize the interaction between biological systems and different types of materials has been investigated, with the ultimate goal of achieving biocompatible surfaces. Cold plasma treatment only affects the surface of the treated material. The physical, chemical, mechanical, electrical, and optical properties of the interior of the material are not altered by the cold plasma. On the other hand, the use of different acids and chemical solvents can damage the surface of many plastics and, if absorbed, affect the properties of the interior of the material.

Some of the biomedical applications of cold plasmas try to improve the adhesion between two surfaces. This work requires the intervention of intense interfacial forces, either through chemical compatibility or by the generation of chemical bonds favored by treatment with cold plasmas.

Among the applications involving an improvement in the adhesion between two surfaces, the pretreatment of catheters, that of components of dialysis pumps, and that of plastics for blood bags or for the packaging of certain drugs, stand out.

Materials that are in contact with blood or proteins require special treatments that improve their biocompatibility. The amino functional groups, obtained from cold ammonia plasmas, act as hooks that retain anticoagulant substances, such as heparin, which reduce the risk of thrombosis. In experimental cell and tissue manipulation protocols, plasma-assisted chemical vapor deposition is already being applied to biodegradable substrates. Treatment with cold plasmas of polystyrene (polymeric material from which the Petri dishes of cell cultures are made) promotes cell adherence and growth. If certain areas of the polystyrene surface are not treated, cells will not adhere to them and, therefore, the structures that would allow the formation of complete biological tissues from seed cells will not be generated.

Cold plasmas are beginning to be seen as offering an alternative method of disinfecting and sterilizing medical supplies. In this sense, cold plasmas would make it possible to simultaneously modify and sterilize the surface of the biomedical material. Plasma sterilization may be suitable in the case of devices sensitive to radiation, to the high temperatures of medical autoclaves, or when aggressive chemicals are involved. The main obstacles against the commercial use of cold plasmas in the biomedical sector derive from the lack of administrative regulation and the scarcity of studies on the reproducibility and validity on larger scales of the biological effects of cold plasmas observed in experiments carried out in academic laboratories. In any case, it appears that the advantages over other methods outweigh the possible disadvantages.

All these applications require the use of atmospheric plasmas. Chapter 7 of this book addresses the generation and applications of high-pressure plasmas in a comprehensive way.

In summary, the medical and biomedical applications of plasmas comprise different subthemes.

- **Sterilization and decontamination by plasma-produced species:** This is the first biomedical application of plasmas. Sterilization of heat-sensitive materials still represents a challenge. Low-temperature sterilization reactors exist, but some of them (like ethylene oxide) produce toxic residuals. By the way, prion sterilization leads to the of use high-treatment temperature (134°C at a steam autoclave). Plasma sterilization could represent a new cold and safe sterilization process. Studies on plasma at a low or atmospheric pressure have shown its efficiency in killing bacteria. Plasmas based on oxygen, nitrogen, or various gas mixtures are tested. Various means of producing plasmas have been found, like microwave and dielectric-barrier discharge (DBD). The antimicrobial action has a variable origin depending on the devices, like ultraviolet radiation, etching action including nitrogen or oxygen atoms produced by the discharge or the postdischarge, and creation of ozone.

- **Surface treatments for medical applications:** This combines biomaterials, tissue development, and bioactive materials. Bioactive materials stimulate a biological response from the body, such as bonding to tissue. Osteoconductive bioactive materials can bond to hard tissue (e.g., bone) and guide bone growth along the surface of the bioactive material. Plasma technology can modify implant surfaces, for example, by covering surfaces with synthetic hydroxyapatite or tricalcium phosphate ceramics. On the other hand, osteoproductive materials stimulate the growth of new bone on the material. It can also stimulate bonding to soft tissue, such as gingival (gum) and cartilage.
- **Therapeutic plasmas for medicine:** Production of plasma at atmospheric pressure (cold atmospheric plasma [CAP]) made possible its use for other medical applications, such as wound healing, blood coagulation, antibacterial treatment, endothelial cell proliferation, and oncology. Some devices are already used clinically; others are still on the bench side.

Two predominant types of plasma-discharge devices can be distinguished: direct discharge sources and indirect discharge sources. Direct plasma discharge sources (e.g., DBD) use the target area as a counterelectrode. These direct plasma sources create relatively homogenous plasmas containing high concentrations of plasma-generated species. Although the control of the plasma composition still remains a big challenge, these direct discharge sources are able to control the plasma composition more easily compared with other discharge devices. The major disadvantage of this technique is the application distance (between the electrodes), which must remain within a close range, generally less than 3 mm^2, thus limiting its use to small areas of the human body. Indirect discharge sources (e.g., plasma jet) refer to various discharge systems used in plasma science. Hence numerous configurations are found. It generally refers to a system where the carrier gas discharge is operated in a nonsealed electrode arrangement. Chapter 7 addresses this topic in depth.

Plasma jets can be classified according to parameters such as discharge geometry, electrode arrangement, excitation, frequency, and pattern. But concentration in reactive oxygen and nitrogen species is lower than in direct discharge sources and the plasma generated seems to be less controllable.

Two methods of applying plasma are also described:

- Direct treatment, which involves directly applying the CAP on in vitro cells, in vivo models, and human living tissues
- Indirect treatment using plasma-activated media or solution

Recently, clinical studies on human patients have been reported, especially in the field of cosmetology and dermatology. This is the proof of the emerging interest of the "medical community" for the potential medical applications of gaseous plasmas.

Chapter 8 describes the state of the art in one of the most popular branches of medicine, odontology.

Chapter 2

Plasma-Enhanced Chemical Vapor Deposition of Thin Films

C. Corbella,[a] **O. Sánchez,**[b] **and J. M. Albella**[b]
[a]*Department of Mechanical & Aerospace Engineering, George Washington University, Washington, DC, USA*
[b]*Institute of Materials Science, CSIC, Cantoblanco, Madrid, Spain*
jmalbella@icmm.csic.es

2.1 Introduction

Chemical vapor deposition (CVD) is a well-established technique for the synthesis of thin film materials on a substrate through the thermal activation of chemical reactions, which take place preferentially on the substrate surface. In conventional CVD techniques, the deposition process is carried out inside a reactor, generally of a tubular geometry, where there is a continuous flow of the reactant gases on the substrate surface.

The gas pressure inside the reactor has a remarkable effect on the homogeneity of the thickness of the deposited layer. According to the fluid dynamics, the transport of gases inside the reactor results in the formation of a boundary (stagnant) layer of the gases

Plasma Applications for Material Modification: From Microelectronics to Biological Materials
Edited by Francisco L. Tabarés

ISBN 978-981-4877-35-0 (Hardcover), 978-1-003-11920-3 (eBook)
www.jennystanford.com

on the substrate surface, where the deposition reaction takes place. Under these conditions, the incoming gases have to diffuse through the boundary layer to reach the substrate surface. However, the diffusivity of gases through this stagnant layer increases inversely to the gas pressure. Therefore, working under low-pressure (LP) conditions (generally < 1 mbar) allows the creation of very conformal coatings, even in 3D architectures of complex geometries. This fact makes possible the production of thin homogeneous layers using LP-CVD techniques of relatively moderate operational costs. These methods are being successfully used in the deposition of a large variety of thin layers, for example, poly-Si, SiO_2, Si_3N_4, TiO_2, and BN, covering multiple applications in diverse fields: metallurgy, solar energy, micro- and optoelectronic devices, biomaterials, etc. On the contrary, operation at higher pressures (i.e., atmospheric) usually promotes lateral reactions in the gas phase, leading to the formation of a powdery product.

Very often, CVD reactions require high temperatures, which may cause damage on substrates not tolerant of high thermal loads. This problem can be minimized through the activation of the gases by other means, namely plasma discharges and laser radiation, working at low or atmospheric pressure (AP). In CVD reactions activated by plasma (PECVD), the transfer of energy to the precursor gas molecules takes place through the acceleration of the electrons generated by the electrical discharge induced inside the reactor. The key steps in these techniques rely on the excitation, ionization, and/or dissociation by electronic impact of the atoms and molecules inside the plasma, thus increasing their reactivity. In this way, the deposit can be made at much lower temperatures (generally between room temperature and some 300°C–400°C) than those used in purely thermal CVD reactors, thus avoiding damage to thermally degradable substrates. Besides, working at LP facilitates the transport of reactant gases to the surface of the sample, which is also beneficial for obtaining good homogeneity in thickness, as stated above. This effect, along with the possibility of generating an extra bombardment with energetic ions from external sources, has led to a wide range of applications of LP techniques (like LP-PECVD).

Nevertheless, when analyzing the operational costs of LP techniques some issues come up related to the limited size of vacuum chambers, which prevents large area coatings as well as operation in continuous manufacturing. Furthermore, in batch processing it is necessary to use complex lock-load chambers to feed the samples into the reactor, raising the operational costs.

To overcome these problems, plasma discharge processes working at AP (such as AP-PECVD) have been thoroughly investigated as an alternative to the LP techniques for film deposition [1, 2]. However, the voltage necessary to produce AP discharges is much higher (generally >10 kV) than those applied at LP (<1 kV) (see Section 2.2). This fact poses serious problems of eliminating spurious arc-discharges including the elevated costs of high voltage sources. Despite this, intense research in this field has led to appropriate designs of plasma sources, allowing the deposition of layers and other surface treatments on large substrates, working in continuous operation lines (even in air discharges). Nonetheless, other limitations of AP techniques, related to the low energy excitation of the gas molecules, have restricted the applications of AP discharges to the deposition of thin films from gas precursors of soft chemical bonds.

This chapter begins with a description of the plasma physics related to the different mechanisms of activation of reactive gases in PECVD techniques under both LP and AP discharges. Due to the different discharge processes and plasma sources in both methods, a comparison between LP and AP techniques of thin film deposition is also included. The paper ends with a discussion of some relevant examples of film deposition of selected materials taken from the literature.

2.2 Effect of Gas Pressure on the Electrical Discharges between Two Electrodes

2.2.1 Paschen's Law

When a gas is subjected to an intense electric field between two parallel electrodes, the primary effect is the distortion and polarization of the gas molecules and eventually the excitation and/or emission of electrons from the outer energy atomic levels of the gas. The electrons emitted, along with those coming from the omnipresent cosmic radiation, are then accelerated by the electric field toward the anode of the discharge. This process may also give rise to new electrons through impact collisions with the gas species, ending up in the production of new excited molecules in the form of ions and radicals, with the corresponding emission of light by de-excitation phenomena. Such effect justifies the term "glow" that characterizes the discharge. Due to the lower mass of electrons as compared to ions and gas molecules, they behave as the main carriers of the kinetic energy supplied by the electric field.

In direct current (DC) discharges, all the limiting walls of the plasma (electrodes, samples, and container) are subjected to a more intense flow of electrons than that of ions because of the higher velocity of electrons in comparison to the heavier ions. Hence, all the walls in contact with the plasma become negatively charged and are surrounded by a "sheath" of positive ions whereas the bulk of the plasma remains practically at a constant voltage. Consequently, most of the drop in the applied voltage, V_B, takes place in a region close to the cathode (known as "cathode fall") associated with the space charge of the positive ions accumulated in this region. Obviously, the sheath thickness at the cathode depends on the gas pressure, P, roughly according to the inverse of the square root of P.

In LP discharges, it is in this sheath, close to the cathode, where most of the power is dissipated by the intense ion bombardment on the cathode surface, releasing new electrons from the cathode through secondary emission processes. Within this area, the electrons emitted from the cathode are accelerated toward the anode, reaching high energies (over hundreds of electron volts). On their way to the anode, these secondary electrons may also give rise to the excitation and ionization of the gas molecules in the plasma bulk through collisional processes, with the consequent formation of electronic avalanches (Fig. 2.1). Under certain conditions, the production of electrons in the cathode can compensate the electrons lost at the anode and container walls; thereby the discharge becomes self-sustaining because of the continuous formation of electronic multiplication processes (known as "Townsend discharge"). As the electric field is practically zero in the bulk of the plasma, the cathode sheath plays an essential role in the transfer of applied electrical energy to the electrons and indirectly to the gas molecules through the collision processes. Alternatively, in AP discharges the ion arrival energy to the cathode is lower than in LP conditions and so is the secondary electron emission. Hence, these phenomena play a minor role in sustaining the discharge, as discussed later.

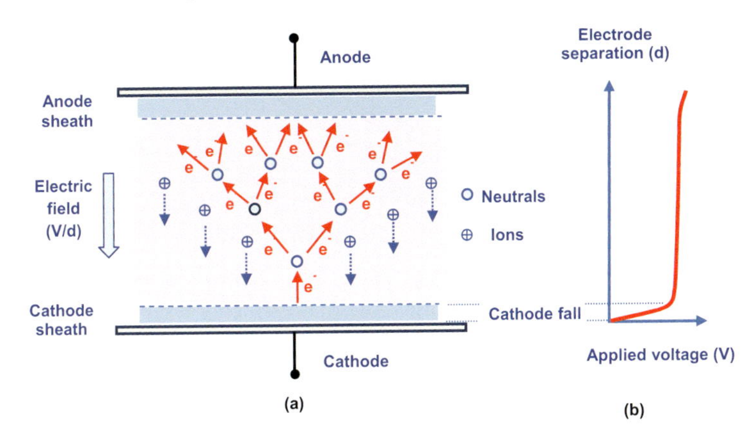

Figure 2.1 (a) Mechanism of the exponential growth of electron avalanches in DC electrical discharges. (b) Qualitative variation of the voltage distribution, *V*, along the distance, *d*, between the electrodes, showing the "cathode fall."

The key point controlling the behavior of the plasma particles under both LP and AP conditions is determined by the well-known empirical Paschen's law, which establishes the relationship between the breakdown voltage, V_B, as a function of the parameter $P \times d$, that is, the product of the reactor pressure times the distance, *d*, between the discharge electrodes. Figure 2.2 shows the experimental curves for some gases (monoatomic and molecular) commonly used in PECVD processes. As can be observed, the curves present a minimum in the breakdown voltage, V_B, increasing toward higher voltages for increasing or decreasing $P \times d$ values with respect to the position of the minimum. By modeling the avalanche processes between the plasma particles, it is possible to adjust the Paschen curves to a theoretical equation of the type

$$V_B = \frac{B \times P \times d}{C + \ln(p \times d)}$$

where *B* and *C* are constant parameters depending on the gas mixture and the cathode material. The law is fulfilled not only in DC but also in a wide frequency range of alternating current (AC) discharges.

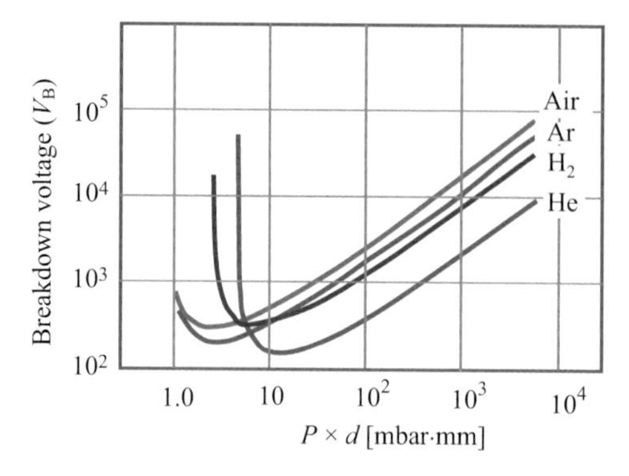

Figure 2.2 Variation in breakdown voltage, V_B, with the product $P \times d$, in electrical discharges for several gases (see the text). Data taken from Merche et al. [3].

This behavior can be explained from a phenomenological point of view. At the minimum of V_B, the discharge is self-sustaining under conditions of the lowest potential drop and power dissipation in the cathode region. Above the minimum, V_B increases as the product $P \times d$ increases because either the average mean free path of the electrons (distance travelled without colliding) becomes smaller as the pressure increases or the electric field decreases as the electrode distance increases. In both cases, the electron energy gained from the electric field between collisions gets smaller. On the other hand, below the minimum, the voltage V_B also increases when the $P \times d$ product decreases because either the number of ionizing collisions becomes smaller at a reduced pressure or the distance between the electrodes is not enough to develop electronic avalanches to produce the gas electrical breakdown. Note that in both branches of the curves, the number of ionizing collisions is smaller than at the minimum.

As can be appreciated in the curves shown in Fig. 2.2, the position of the minimum, $V_{B,min}$, depends on the gas used in the discharge. Among other factors, this effect is associated with differences between either the ionization potentials of the gases or the mean lifetimes of their excited metastable states. Helium, despite its large ionization energy (24.5 eV), is the gas with a lower $V_{B,min}$, with the corresponding product $P \times d$ being very high and, besides, it keeps more stable discharges. In this case, the plasma is sustained by the excitation of He atoms to a metastable state of low energy

(~20 eV), but with a long lifetime, thus contributing to the ionization of other species through Penning collisions (see Section 2.3.1). For this reason, He as well as other gases (Ar, N_2, etc.) are commonly used as diluents in processes at AP. Importantly, air shows one of the largest breakdown voltages and its use is not recommended. In particular, an interelectrode gap of 1 cm requires the application of 30 kV to produce a DC discharge in air under AP conditions. On the other hand, at LP ($10^{-5} < P < 10^{-3}$ mbar) under similar conditions, the breaking voltage is lower than 10^3 V, which represents a considerable advantage of the LP-PECVD techniques over the AP-PECVD techniques [3–5].

2.2.2 Thermal and Low-Temperature Plasma Discharges

Generally, plasmas are classified as low-temperature (cold) plasmas and high-temperature or thermal (hot) plasmas. In thermal plasmas, the ions and electrons may achieve very high temperatures so the gas can reach complete ionization. In this case, the main applications are in fusion research using a plasma with a high electron density (around 10^{19} cm^{-3}), whereas the electron equivalent temperatures are in the range of 10^6 K. In contrast, cold plasmas are often only partially ionized (typically, 10^8–10^{13} cm^{-3}). The temperature of the heavy particles, that is, ions and neutrals, generally shows values near room temperature, while only the electrons have the high energy required for ionization and dissociation processes. These processes form the basis of a large number of surface treatments in the semiconductor industry, such as plasma etching and plasma deposition. Roughly, in all these processes the reactive gas precursor is dissociated by the absorption of electrical energy. The dissociation products may react either with one another or with the substrate, leading to material deposition (film growth) or material removal (etching), as stated above.

The gas pressure in the discharge has a strong effect on the energy of the particles in the plasma due to different collision processes that can take place between them. In the case of LP (≤1 mbar), that is, low gas density, the collision frequency of the electrons with the other particles is very low and their mean free path is very large (0.1 m). Under these conditions, the electrons are accelerated by the intense electric field near the sheaths to high energies, enough to provoke the excitation and ionization of the neutral gas. Thereafter, the colliding electrons lose most of their energy, though they can be reaccelerated before a new collision event takes place. Such reacceleration processes of electrons may be promoted by potential gradients located at the sheath-edge region. As a result of the energy gain from the electric field, the energy distribution of electrons departs from the typical Maxwellian curve for electrons in thermal equilibrium, showing a tail on the high energy side (see Fig. 2.4). Deviation from the Maxwellian distribution is observed especially at high values of the electric field, where the energetic electrons can either be depleted due to ionization or dissociation processes or be generated in virtue of "heating processes" by the strong electric field.

Under these circumstances, a unique Maxwell distribution cannot be applied for both electrons and gas species in LP plasmas, since every ensemble of particles, namely electrons and heavy species like ions and neutrals, is moving collectively in different ranges of energy. In the case of electrons, the deviation from the theoretical Maxwell curve has led to the definition of an "equivalent temperature" (T_e) for a narrow electric field interval, which can be much higher than that of the gas molecules whose kinetic energy gain is limited by excitation mechanisms (rotation or vibration temperatures). The plasma is said to be in nonlocal thermal equilibrium (non-LTE) between electrons and gas particles, where T_e can vary in the range of 10^4–10^5 K, whereas the gas temperature generally oscillates between 300 K and 1000 K.

On the other hand, at higher pressures (≥1 mbar) the collision frequency between electrons and neutrals is very high (around 10^{10}–10^{12} s^{-1}), causing the electrons to be rapidly thermalized with the gas. Consequently, both temperatures—that of electrons and that of gas molecules—are balanced at about 10^4 K. Importantly, the applied power density per gas molecule to induce the discharge strongly influences the plasma state [6]. Generally, high power densities favor LTE conditions; an example is the arc discharges. The natural tendency is also to approach global thermal equilibrium at AP. Actually, when the pressure is increased, the gas temperature increases smoothly because in this case, energy losses by diffusion within the plasma volume are less important.

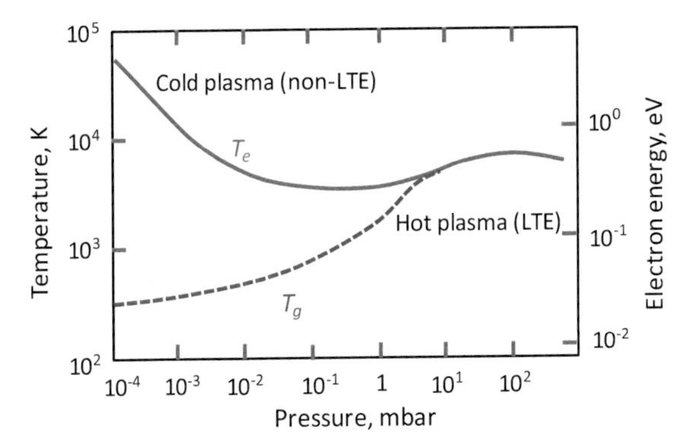

Figure 2.3 Typical dependence of the electron temperature (T_e) and the gas temperature (T_g) on the discharge pressure.

To summarize, when the gas pressure increases from LP to AP conditions, the final effect is a gradual transition from a non-local thermal energy distribution among the different plasma species (non-LTE) to a unique or global thermal equilibrium for all of them (LTE). Figure 2.3 sketches the typical variations in the electron and gas temperatures with the discharge gas pressure in a typical Ar plasma discharge. At LP (≤1 mbar) the temperature of the electrons is much higher than that of the gas. Depending on the gas mixture, both temperatures may merge at about 10 mbar.

Regardless of the effect of pressure, other plasma conditions, such as electrode gap, presence of magnetic fields, discharge volume, and application of either AC or high voltage pulses, may lead to non-LTE conditions, even at AP. As discussed below, these procedures have made possible the development of a vast variety of discharge configurations working at high pressures, such as the so-called dielectric barrier discharges (DBD) and the inductively coupled plasmas.

2.3 Elementary Collisional Processes in Plasma Discharges

2.3.1 Elastic and Inelastic Collision Processes

In plasma discharges between two electrodes, excitation of species and transport phenomena of electrons in the gas phase are crucial for the creation and sustenance of the plasma. As stated earlier, the electrons absorb kinetic energy as a consequence of the high electric field at the cathode, giving rise to ionization events in the bulk plasma. The acceleration of electrons and energy gain depend on their collision rate with the background gas and hence on the discharge pressure. Important processes in plasmas are the scattering events of charged particles with each other as well as ionization processes. The ionization degree of plasmas is determined by the balance between the creation and annihilation rates of ions.

Energy transfer among gas particles takes place by these interacting collisions. In particular, the energy acquired by the electrons in the plasma is partially lost through elastic collisions by Coulomb interactions of the electrons with neutrals and charged particles of the plasma, all of them becoming thermalized in AP discharges, where the collision frequency is higher. More complex processes, such as the excitation of chemical species, the formation of radicals, and recombination, are also ruled by the frequency of inelastic collisions, where changes in the internal energy and restructuring of the involved species take place. These processes in the gas phase are very relevant in plasmas and, particularly, in PECVD techniques, since they increase the reactivity of the precursor gases. Due to the smaller mass of electrons in comparison to that of atoms or molecules, the kinetic energy transferred by energetic electrons to the heavy particles in elastic collisions is very low, about a factor of 10^{-4} of their initial kinetic energy.

Collision processes are generally described by means of a cross section, which represents the probability of particle interaction with a partner within a gas of density n_g. In a first approach, the collision cross section, σ, is related to the effective geometrical area of interaction between two colliding particles of the gas. Once σ is known, the mean free path, λ, of the scattering particles can be obtained through the inverse relation

$$\lambda = 1/n_g\sigma$$

Further parameters can be defined. For instance, the mean collision period τ for a particle with velocity u is $\tau = \lambda/u$, which leads to the definition of collision frequency, ν. Clearly, this frequency increases with gas pressure according to

$$\nu = un_g\sigma$$

The typical collision frequency values between electrons and heavy species are in the range of 10^7–10^9 s^{-1} for LP, whereas for AP conditions, they are much higher, 10^{10}–10^{12} s^{-1}. Such values are estimated by considering typical collision cross sections of approximately 10^{-16} cm^2.

The associated cross sections of elastic and inelastic collisions depend strongly on the electron energy, showing a maximum in both cases (Fig. 2.4). Above the maximum, both σ curves decrease since in this range, the electrons reach high velocities, thus decreasing the interaction and the transfer of energy to gas particles. Moreover, both functions appear energy-shifted because inelastic collisions require a minimum threshold energy of several electron volts to activate the excited states of the gas species. And more importantly, a large fraction of the electron energy distribution function, $f(E)$, overlaps with the cross-section curve of elastic collisions, which is in contrast to the small overlapping area with the inelastic collision curve. It means that only the electrons in the high-energy tail of $f(E)$ can contribute to elastic and, to a lesser extent, inelastic processes.

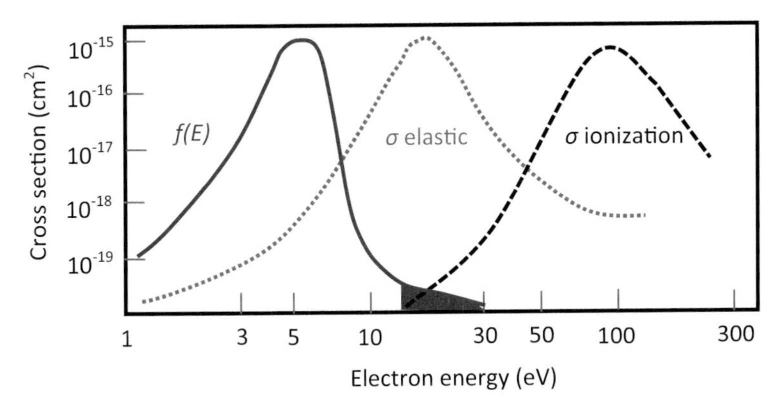

Figure 2.4 Typical curves of the electron cross section of elastic and inelastic collisions for Ar in a glow discharge. For comparative purposes the qualitative electron energy distribution function, $f(E)$, is also included. The dashed area represents the fraction of high-energy electrons able to ionize Ar atoms (ionization energy = 15.8 eV).

Two distinct elastic collisions, where only the orientation and module of the electron velocity are changed, have been observed: (i) Coulomb scattering, associated with the screening of the

Coulomb potential of the ions by the surrounding electrons in the gas phase and (ii) polarization (Langevin) scattering, arising from the interaction of electrons colliding with neutral particles. In this case, the electron shell of the neutral particle becomes polarized as an effect of the electric field of the incident electron. Such processes lead to a thermalization of the whole system.

Inelastic collisions are characterized by a partial transfer of the electron energy to the excitation of the gas species. Such interactions are responsible, among other processes, for the increase of rotational and vibrational temperatures of the gas molecules, including the formation of radicals, thereby increasing their reactivity. The main contributions to inelastic collisions can be summarized as:

- Excitation: After collision with energetic electrons, atoms or molecules can be excited to higher energy levels (rotational, vibrational, or electronic), according to the reaction

  $$e_{\text{high energy}} + A \rightarrow A^{*} + e_{\text{low energy}}.$$

  In the case of molecular gases, these types of collisions may activate the production of the reactive species needed in the synthesis of new materials. The excited atom or molecule may return to the ground state through photon emission (typical glow of the discharges) whose energy is determined by the difference of the initial and final levels. Alternatively, the atom may return to a metastable state, with a longer lifetime (in the range of 10^{-3} s), retaining its energy for new collisional processes. In fact, this is the only allowed path to return to the original ground level.
- Ionization: Electron-impact ionization takes place only when the electron energy is higher than the ionization potential. Above this energy threshold, the corresponding cross section increases strongly up to a maximum, which is around 70 eV for most atoms and molecules (Fig. 2.4). Stepwise ionization may also take place below the energy threshold when the previously excited neutral atom, A^{*}, is in a metastable state, for example

  $$e_{\text{high energy}} + A \rightarrow e_{\text{low energy}} + A^{*} \rightarrow e_{\text{low energy}} + A^{+} + e$$

  The impact ionization coefficient, α, (also known as first Townsend coefficient) determines the probability that an electron emitted from the cathode collides with a neutral gas particle along its way to the cathode, forming an ion plus an additional electron. It is defined as the number of ionizing collisions per unit length, which is obviously proportional to the gas pressure and, therefore, to the inverse of the electron mean free path. These ionizing collisions are the source of new electrons and ions, which substantially contribute to the plasma sustaining.

- Recombination: This process may occur when an electron collides with an ion that passes into an excited state and eventually may emit a photon (radiative recombination):

$$e + A^{+} \rightarrow A^{*} \rightarrow A + h\nu$$

Such reactions take place preferentially at higher pressures, since two-particle recombination requires the participation of a third particle (generally an additional electron), which balances the momentum and energy of the system.

- Electron capture: Plasma electrons can be captured by atoms and/or molecules to form negative ions. However, here the associated cross section, which is higher for elements with high electron affinity atoms, is very small. Electrons must have energies close to the ionizing energy for a successful attachment reaction

$$e + A \rightarrow A^{-}$$

- Penning excitation: Atomic or molecular species highly excited by previous collisions are capable of exciting and/or even ionizing other species initially neutral after new collisions:

$$A^{*} + B \rightarrow A + B^{*}$$

This reaction of energy transfer is of upmost importance in atmospheric plasmas, where the electron' mean free path is very low and consequently their kinetic energy is not enough to cause gas ionization. The Penning process is advantageously used to decrease the minimum breakdown voltage of some gas mixtures, as it occurs in He discharges.

- Molecular dissociation: Electrons whose kinetic energy is higher than the dissociation energy of the molecules can break the atomic bonds of molecules and release single atoms or radicals:

$$e + AB \rightarrow A + B + e$$

- Charge-exchange collisions: Positive ions and atoms may exchange charge states after transfer of a valence electron from the atom to the ion:

$$A^+ + B \rightarrow A + B^+$$

- Combination reaction: Under the presence of reactive species in a plasma discharge, chemical reactions activated by the discharge may provide new compounds according to the overall equation

$$AB + CD \rightarrow AC + BD$$

  This is the basis of PECVD techniques, where the plasma discharge allows obtaining deposits of new compounds at much lower temperatures than the corresponding CVD processes. The detailed reaction kinetics is much more complex than the preceding elementary reactions.

Besides the preceding collision reactions, it is worth taking into account secondary emission processes at the cathode of the discharge produced by the bombarding ions in LP plasmas, which in the case of DC is the most important electron source sustaining the plasma. In this process, there is a threshold of the ion bombarding energy needed for electron emission, which should be about twice that of the electronic work function of the cathode (i.e., the energy required to extract an electron from the Fermi level of the material). The energy of the emitted electrons varies in a broad range, with a maximum between 2 and 6 eV, regardless of the energy of the incident ion. The electron yield, that is, the mean number of electrons emitted per incident ion (denoted by γ) is relatively low ($\gamma = 0.02 - 0.15$) for Ar^+. It depends on the material (metal or semiconductor), its structure and surface morphology, the presence of impurities, etc.

Apart from the preceding processes, when the bombarding energy of the ions in the discharge is high enough (some hundreds of electron volts), these ions can also inflict considerable damage on the cathode, particularly by the ejection of surface atoms, which may travel through the plasma to be deposited on a dedicated substrate. This is the basis of the well-known "sputtering" processes, used mostly in the deposition of thin films. The sputtering yield (the number of atoms ejected by one incoming ion) depends obviously on the mass and energy of the bombarding ions (generally argon) as well as on the bonding energy between the atoms in the cathode (target).

2.3.2 Effect of the Discharge Frequency on the Collision Processes

Plasma capacitive reactors allow relatively simple designs, and they are applied in the processing of materials, either for surface modification or for film deposition. They can be used under DC or AC polarization. However, one of the most important limitations in DC discharges is the high fraction of energy dissipated thermally by the heat arising from the intense bombardment of the ions on the cathode. However, only a low percentage of the electric energy (approximately 10%) is spent in plasma activation or heating. Generally, AC electrical discharges, particularly in the high-frequency domain, are much more efficient in atom and molecule ionization than the DC counterparts.

Different ways in which the driven power energy is coupled with the plasma can be discussed in terms of the so-called natural or resonance frequency of the collective oscillations of charged particles, that is, ions, ω_i, and electrons (termed as "plasmons"), ω_e, which determine how they react to changes in the frequency of the electric field. Below their resonance frequency, the charged particles (ion and electrons) can follow the field oscillations, while above that frequency, they move statistically in all directions. Obviously, due to the larger mass of the ions, $\omega_i << \omega_e$, with ω_i in the range of 100 kHz.

2.3.2.1 Discharges at the DC-kHz regimes ($\omega_{kHz} < \omega_i < \omega_e$)

Glow discharges operated with DC as well as with AC up to the kHz region show similar behavior. In the latter case (AC), the electrodes behave alternatively as anode and cathode of the discharge. The reason is that all species, light and heavily charged, can follow the electric field oscillations. In such discharges, charge carrier conservation requires the generation of secondary electrons at the electrode level, as stated above [7]. In this regime, the current flow sets a limit for the application of such plasmas to nonconductive electrodes. Insulating electrodes or ceramic-deposited films, such as oxides or oxynitrides, leads to charge build-up effects, which might result in arcing. This charging up effect can be avoided by increasing the excitation frequency to promote charge compensation at each semicycle. However, thermal emission takes place when very high currents flow to the electrodes, ending up in arc discharges [8]. These "hot" (energetic) discharges are possible both in LP and AP plasmas. Such plasmas are thermally sustained due to the high gas temperatures (0.5–1 eV) achieved in the arc plasma core. Indeed, they are basically sustained by the thermionic emission of electrons from a hot cathode surface as well as by the generation and acceleration of ions taking place preferentially at the anode sheath.

The DC and kHz-regime plasmas have a wide scope of applications due to the cost-effective power supply setups. Required voltages are usually within the range of some kilovolts, thus promoting intense ion bombardment onto deposited films. The energy distribution of the incident ions on a biased cathode has a strong influence on the final coating properties. Indeed, surface properties of the growing films are strongly correlated to the flux and energy of incident particles from the plasma, whose values are determined by plasma density and sheath properties. For example, energetic ion influxes are recurrent in the deposition of dense and hard coatings, thanks to the generation of a strong atomic network in the films. In contrast, deposition processes with lower ion energies tend to produce soft, polymer-like materials owing to a more relaxed structure [9].

2.3.2.2 Discharges at the MHz regime ($\omega_i < \omega_{rf} < \omega_e$)

The frequency should be high enough to avoid charging up effects of the insulating layers deposited on the electrodes (typically 13.56 MHz). In this range, the excitation frequency in such plasmas, ω_{rf}, is higher than the ion resonance plasma frequency but lower than the frequency of electrons. Roughly, ions cannot follow the oscillations of the electric field as electrons do, and so the ions show a different behavior from the electrons, which implies non-LTE conditions. In particular, the higher mobility of electrons as compared with ions may give rise to ionizing collisions even in atmospheric plasmas, thus allowing large area processing [10]. In capacitive discharges, heating of electrons by means of a displacement current permits efficient powering of electrodes, even with insulating surfaces. Nevertheless, covering one or both electrodes with a dielectric material gives more stable AP discharges (see Section 2.4). Since the net current toward surfaces is now negligible, the charging up effects on the coated substrates vanish. The applied voltages are on the order of several 100 V up to some kilovolts.

In LP-PECVD applications, reactors are designed according to capacitively coupled plasma (CCP) discharges. Typical configurations accept electrode setups with different degrees of symmetry. Usually, asymmetric electrode areas are selected; generally, the largest one is grounded (connected to the chamber) whereas the smaller electrode is driven by the applied radio frequency (RF) power. In this technique, it is necessary to insert a capacitor of large capacitance between the RF power supply and the small electrode. After the initial oscillation periods of the applied AC voltage, the capacitor becomes fully charged, adding a negative constant voltage to the small electrode (self-bias polarization). Under these conditions, this electrode acts as a cathode of the discharge, being subjected to an intense bombardment of the discharge ions (hundreds of electron volts), in a way similar to DC glow discharges (Fig. 2.5). This configuration is common in thin film deposition as well as in plasma etching [11].

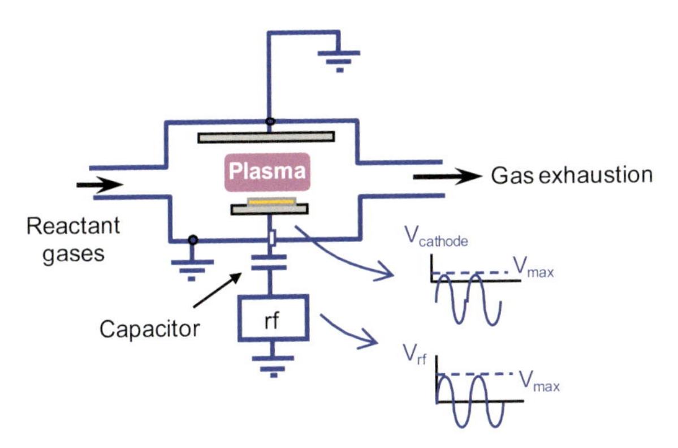

Figure 2.5 Sketch of a CCP reactor for film deposition working in the mid-RF range, in the self-bias configuration used in LP-PECVD techniques.

Depending on the applied pressure and voltage, in this configuration, CCP discharges can operate in two distinct regimes relevant for sustaining the plasma: stochastic heating and ohmic heating. At relatively moderate pressures (about 1 mbar) and RF voltages (≈1 kV), stochastic heating is mainly produced by the collision of electrons with sheath edges that oscillate due to the RF field. Such a process, together with the oscillations of said electrons to follow the RF field, constitutes an efficient heating mechanism that dominates at LP or noncollisional plasmas, whose characteristic mean free path is similar to or larger than the CCP reactor dimensions. In such mode, termed as "γ mode," ionization events take place preferentially at the plasma boundary region. Instead, for higher pressures (>100 mbar) and voltages, collisional processes within the plasma bulk dominate over stochastic heating via sheath oscillations. This mode is known as ohmic heating (α mode) [5, 12–14].

As mentioned above, the properties of films deposited at LP in a plasma discharge can be tailored by controlling the energy of the incident ions on the growing film surface. However, this control requires plasmas with a high ionization degree because only charged particles can be manipulated by electric and/or magnetic fields. This plasma state of almost total ionization can be achieved by alternative physical deposition methods (PVD), which cannot be dealt with here in detail. For example, in sputtering deposition techniques, the cathode can be driven with very short pulses (≈100 μs) and high voltages (1–4 kV) in the so-called high-power impulse magnetron sputtering (HiPIMS) [15]. Excitation frequencies of several tens of Hertz and very short duty cycles (ca. <1%) provide peak currents of up to 1000 A during the pulse, that is, peak powers in the megawatt range. Recently, diamond-like carbon (DLC) films with improved mechanical properties have been produced by HiPIMS and PECVD using pulsed-DC sources [16, 17].

2.3.2.3 Discharges at the GHz regime ($\omega_i < \omega_{rf} \approx \omega_e$)

Similar to MHz-regime discharges, in this regime, the ions cannot follow the variations in the electric field in plasmas operating in the range of microwaves. In this case, electrons follow, only partially, the rapidly changing electric fields. The refractive index in the plasma is almost unity in this frequency domain, meaning that the plasma becomes transparent for incident electromagnetic waves [12]. In fact, the electric permittivity, ε_p, of nonmagnetized plasmas can be expressed as

$$\varepsilon_p = \varepsilon_0 \left(1 - \omega_{pe}^2/\omega^2\right)$$

From this expression, it is evident that ε_p is positive and tends to the dielectric vacuum constant, ε_0, at driving frequencies ω much higher than the electron plasma frequency ω_{pe} (high microwave regime). This weak interaction with the plasma contrasts with the situation at low excitation frequencies, where $\varepsilon_p < 0$ and therefore the radiation is blocked by electrostatic screening of the plasma.

Electron heating in this case is mainly performed by the displacement current but not by stochastic heating. The latter heating mechanism is not effective here because the absolute electric field at the powered electrode is smaller compared to MHz discharges. This effect can be explained by the lower energy dissipation observed in plasma sheaths at very high frequencies, so high that sheath edges cannot follow the excitation frequency. However, resonant heating by other means is now possible by the application of a 100 mT parallel magnetic field. This configuration gives rise to synchronous resonant oscillations of the electrons following the electric field, along with the rotation around the magnetic field lines. This additional electron heating mechanism enhances substantially the electron density (about 10^{11}–10^{13} cm^{-3}), which allows working at lower pressures (<1 mbar), thus increasing the electron mean free path and consequently their energy. As a consequence, the incident flux of charged species toward the substrate is enhanced during film deposition. It is worth noting that this type of microwave discharge allows independent control of ion energy and current, as opposed to the conventional CCP sources. Actually, the magnetically aided discharges have led to the development of new LP high-density electron sources with different configurations (helicon), resulting in a large variety of technical applications [11, 18].

In the plasma remote configuration for the deposition of thin films, a second magnetic coil is added to the chamber so that the divergent magnetic field drags the charged molecular species toward a second chamber placed at the bottom, where they are deposited (Fig. 2.6). This configuration avoids damaging the samples from the plasma radiation. Besides, it is possible to add a negative bias to the substrates in order to increase the energy with which the plasma ions bombard the growing film surface.

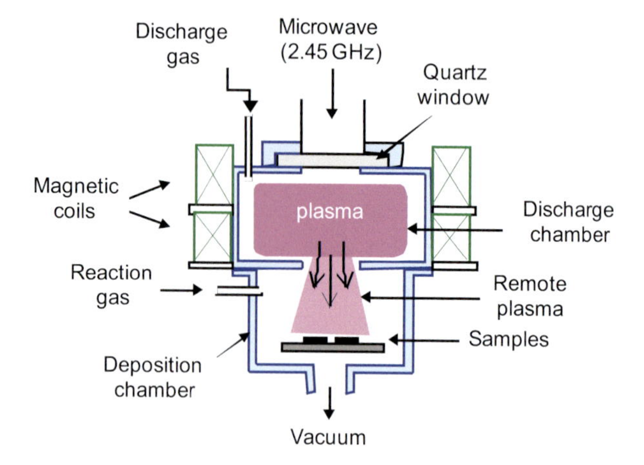

Figure 2.6 Sketch of an ECR reactor for film deposition, provided with two magnetic Tesla coils, to get plasma resonant conditions (upper coil) and to extract the discharge toward the substrate (lower coil).

Additional magnetrons are usually installed in GHz reactors to heat the plasma by this electron cyclotron resonance (ECR) effect. The amplitude of the electric field is under 100 V. However, the supplied power in this case is absorbed locally, which bears an important issue in GHz plasmas, since the chamber volume occupied by the plasma is very limited. Besides, the relatively short wavelengths (in centimeters) promote the formation of standing waves in the plasma chamber, which makes the deposition of homogeneous films challenging in GHz discharges. Lack of plasma homogeneity due to interference patterns of the fields is also a common issue in MHz discharges, especially in large-scale reactors [19, 20].

Plasma generation by means of two or more independent power sources with different frequencies (multifrequency plasma) is also a well-known technique to improve control over depositing film properties. The high-frequency component (in the microwave range) excites the glow discharge and determines the electron density, whereas a low frequency (RF) component controls the bias voltage, thus modulating the ion bombardment onto the substrate. This method permits the decoupling of the electron density of the plasma and the energy of the incident ions [2, 21].

2.4 LP-PECVD vs. AP-PECVD of Thin Films

In the past 50 years, LP-CVD, particularly LP-PECVD techniques, have reached a high degree of maturity, mostly fostered by their vast applications in the microelectronic industry, which requires strict control of the properties of the deposited layers (purity near the limit of parts per billion, epitaxial crystallinity, thickness homogeneity in the nanometric range, etc.). These techniques were recently successfully transferred to other applications in the emerging field of "functional" coatings (mechanical and corrosion protection, thermal barriers, optical and magnetic devices, decorative coatings, solar energy harvesting, biomaterials, etc.). The past few years have witnessed a lot of progress, accompanied by abundant bibliography on this topic. However, the use of complex high-vacuum systems, with their inherent costs in the deposition of films, including long processing times, poses serious limitations for in-line deposition of large area substrates. Thus, these LP techniques have been mainly used in the fabrication of samples of reduced dimensions in batch production.

Recent advances in the development of AP discharges, including specially designed plasma sources, have renewed the interest in the scientific community to transfer the knowledge acquired in film deposition techniques under LP conditions to those operating at medium or high (atmospheric) pressure. Within this scenario, a wide number of specially designed electrode configurations have emerged to overcome the difficulties associated with the high voltage needed to ignite the AP discharges as well as to increase the electron energy needed to activate plasma reactions (DBD, hollow cathode, corona, inductively coupled, torch and jet discharges, etc.). The efforts have been addressed to emulate "non-LTE" discharges with their inherent advantages, that is, relatively high electron temperatures leading to inelastic collisions, with the formation of highly reactive species (free radicals, ions and excited atoms) for film deposition and surface modification. Fortunately, despite the considerable differences between LP and AP sources and their operating conditions, the elementary collision processes are indeed the same in both cases.

Within this frame, several approaches have been tried to generate non-LTE plasmas under AP conditions, requiring high-energy electrons whereas the gas molecules remain "cold." Essentially, a low gas temperature can be achieved in two different ways: (i) by limiting the energy applied in the discharge, either by increasing the gas flow in the reactor or by reducing gas transit through the discharge zone and (ii) by direct cooling of the discharge chamber, electrodes, and sidewalls. Moreover, the higher mobility of electrons compared to heavy ions makes necessary the application of medium-high frequency AC voltages (in the kHz–MHz range, depending on the technique) in order to "pump" selectively the electric field energy toward the electrons [1].

Nevertheless, at high pressures, the electron mean free path is very short (<1 μm), so the electron energy between collisions falls typically in the 0.1–2 eV range, which limits the applications of the AP discharges only to the activation of chemical reactions by Penning collisions, with the formation of excited radical species, excluding ionization processes [22]. This may lead also to reduced surface mobility of the chemical species, resulting in amorphous and low-density films. Therefore, the absence of ion bombardment processes demands in some cases additional thermal annealing treatments to obtain dense crystalline films. In this regard, AP techniques cannot compete with those using LP conditions. Whatever the case, it is important to keep in mind that the kinetics and chemical reactions paths in the synthesis of compounds by LPCVD (LP or AP) may differ substantially from those followed by thermal activation (CVD), frequently giving rise to unwanted incorporation of chemical species in the films.

Anyway, due to the large variety of non-equilibrium reactions and electrode configurations, as well as of the discharge parameters (temperature, gas flow rate, applied voltage and frequency, etc.), a great number of AP techniques have been proposed. Most of them are addressed to cover a large variety of applications, other than thin film deposition, such as surface treatments (e.g., to increase the surface energy of polymers or wettability of fabrics), sterilization, synthesis of powders, chemical attack, polymerization reactions, abatement of volatile organic compound, fuel reforming, analytical chemistry, and TV plasma screens. Actually, this motivation has increased the development of new AP techniques and tools, surpassing most traditional LP discharges.

Other advantages of AP reactors are the possibility to work in open systems, in continuous processing of large surfaces, in some cases with remote sources, thus avoiding the elevated costs of vacuum chambers with complex load-lock ports. However, working at high pressures leads to some issues still not resolved, such as poor thickness uniformity of the deposited layers on 3D substrates, including the formation of powders, which are drawbacks inherent to AP-CVD techniques. Moreover, the need for the dilution of the precursor gases in more expensive gases (He or other inert gases) may also challenge production costs.

In addition, the physics of cold (low-temperature) AP plasmas is still under development. First, the transfer of LP results to cold AP requires adapting reactor dimensions to the new plasma physics. Moreover, the focus of research is shifted from ion dynamics, which is relevant in LP plasmas, to the plasma chemistry and to the role of UV radiation. Indeed, energetic ions (>1 eV) do not reach electrodes in AP plasmas due to the high collision rate with the plasma sheaths, so their role is associated with chemical interaction with the surface. In this situation, excited metastables, fast neutrals, reactive agents, and UV photons are relevant species besides plasma ions.

Referring specifically to thin film deposition, one of most used configurations in AP discharges for this purpose is the aforementioned DBD source, initially developed for ozone production. The simplest plasma source is basically designed with two parallel electrodes, one or both covered with a dielectric material (glass, quartz, ceramics, or polymers), leaving a small gap in between where the plasma is sustained, with the substrate placed on one of the electrodes. Figure 2.7 shows some basic configurations [23]. The charge collected at the dielectric surface reduces the voltage drop along the gap and prevents arcing. Obviously, the dielectric should have very good insulating properties for the applied voltage (i.e., high DC resistivity), as well as high breakdown strength. Within these limitations, the electrode configuration of DBD sources is similar to that of the RF-CCP discharges already described, although the operation conditions and discharge processes are quite different.

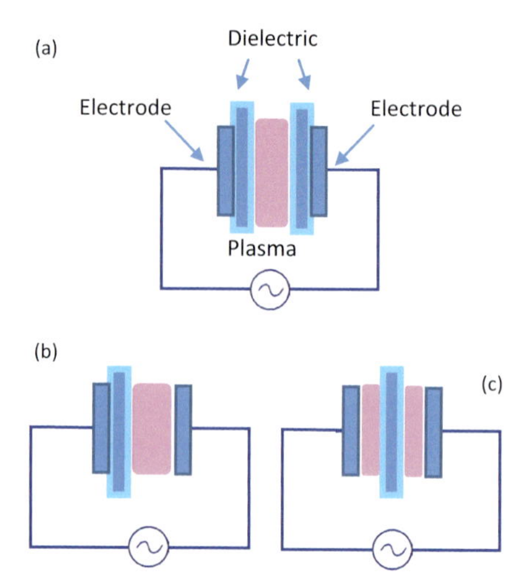

Figure 2.7 Sketch of some basic DBD configurations: (a) symmetric, (b) symmetric, and (c) floated dielectric.

In particular, to avoid the use of high-voltage sources needed to ignite the discharge in DBD, it is necessary to reduce the gap in the range of 0.1–10 mm, including the application of AC power, with a voltage amplitude high enough to provoke the plasma discharge (about 5–15 kV) under non-LTE conditions. The presence of the dielectric implies the use of either low-frequency AC (typically 50 Hz to 500 kHz) or square pulsed discharges to limit the formation of high displacement currents, which are determined by the time derivative of the applied voltage [$\mathrm{d}V(t)/\mathrm{d}t$], as well as by the thickness and dielectric constant of the insulating material. Advantageously, AP discharges working with AC generally show lower values of the minimum voltage in the Paschen curves ($V_{\mathrm{B,min}}$) [1, 4].

The elementary processes underlying DBD discharges are complex. Currently, this topic is still subjected to intense research and debate. Depending on the applied energy and gas pressure, two major distinct regimes of DBD discharges can be distinguished [5]:

- At high pressures (particularly, for $P \times d > 200$ mbar cm), the discharge is initiated in the form of individual filaments (FDBD), typical of capacitive discharges in air. They are characterized by local Townsend avalanches in the plasma bulk, with a large impact ionization coefficient, α. In this mode, the secondary electron emission at the cathode plays a secondary role. The charge separation of electrons and ions in their course to the electrodes creates an additional spatial electric field enhancing the growth of additional avalanches. Consequently, the discharge is localized in distinct channels of small radii (0.2–0.4 mm), randomly distributed across the dielectric area, bridging the cathode and the anode. This is the so-called "streamer" or "filamentary" discharge mode.

- Alternatively, lower $P \times d$ values (<30 mbar cm) give rise to a uniform or diffuse "glow discharge" (GDBD) covering the whole electrode area. They are observed typically in pure inert gases (particularly He) with some chemical precursors diluted, whereas the filamentary discharges are more frequent in molecular gases. The gas flow rate, the applied voltage and waveform, the dielectric material, and the reactor geometry determine the discharge mode as well.

The streamer mode, in the form of microdischarges, usually of 10 ns duration and randomly distributed between both electrodes, results in non-uniform coatings. In addition, the presence of some type of protuberances may cause thickness in-homogeneities in the film deposited. There are several options to prevent the formation of streamers—the use either of He in the discharge or of dielectrics limiting the current, including electrodes with special shapes.

2.5 PECVD of Thin Films under LP and AP Conditions: Some Examples

In the following, we present some examples of plasma processes typically used in the deposition of thin film layers, widely used in technological applications. PECVD permits the growth of thin film materials, in AP and LP conditions, with tunable physical and chemical properties by adequate adjustment of the deposition parameters. Actually, the elementary processes in LP-PECVD are very similar to those observed in conventional LP-CVD, although the reaction paths and the intermediate species may be different, and hence the final composition and structure of the films.

Whatever the case, one can tailor the deposited coatings depending on the desired final application, such as materials with high electric resistivity, high optical transmittance, and low wettability. Just for comparison purposes between LP and AP deposition techniques, we have selected CCP and DBD discharges, which have similar electrode configurations. In particular, DBD discharges are ideal for plasma synthesis and/or treatment of "soft" materials, particularly in the field of polymer-like materials (organic and inorganic), even in open-air systems. In most cases, the electron energies fall in the 1–10 eV range, which is enough for the excitation of chemical species with the rupture of bonds, such as C–C, C–H, and C–O, present in organic precursors. This allows much lower temperatures for film deposition than those required in the absence of the plasma [10, 24, 25]. With this in mind, this section is focused on a limited range of relevant coating materials, namely C, Si, and Ti compounds, appearing in a large variety of applications.

2.5.1 Carbon-Based Compounds

Traditionally, hard ceramic films have been successfully deposited by PECVD. This technique provides many advantages, like adjustable ion bombardment and low substrate temperature, which are beneficial for obtaining hard and well-adhered films. For example, hydrogenated amorphous carbon (a-C:H) with diamond-like properties, commonly known as DLC, has a long record of achievements due to the attractive combination of its surface properties—high hardness, low friction coefficient, and high infrared (IR) transparency—which justify its performance as an excellent tribological coating (machining tools, hard drives, car engine components, etc.) [26, 27]. This material is also well known for its unique biocompatible properties (implants, stents, etc.) [28]. Common precursors to grow DLC by LP-PECVD are hydrocarbon gases, such as CH_4 and C_2H_2. However, the substrate adherence of DLC films is usually hindered by their high compressive stress. Typical solutions consist in the deposition of intermediate adhesion layers and/or combination with other elements, such as metals (e.g., Cr), which reduce stress by introducing a nanocomposite structure. Chromium is a carbide former, able to stabilize the structure and relax internal stress in carbon films [29]. The inclusion of new elements in the a-C:H network can be performed by reactive magnetron sputtering of metal targets, injecting metal-organic precursors [30], directly spreading metal nanoparticles onto the substrate followed by a-C:H deposition [31].

Figure 2.8 shows a ternary-phase diagram of DLC compounds and their structure. Switching from polymer-like to diamond-like properties (inherited from the carbon sp^3/sp^2 hybridization fraction and hydrogen content) can be easily achieved by changing the supplied power, substrate bias, and H_2 concentration in the PECVD discharge [21, 32]. More recent research is focused on adapting DLC deposition on elastic substrates, where achievement of maximal adhesion is the main challenge [33].

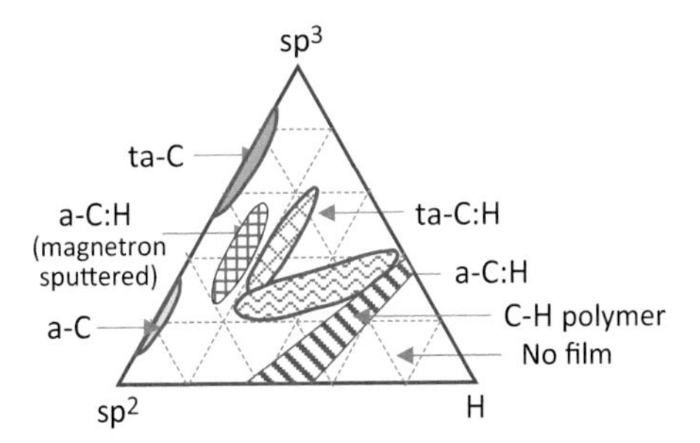

Figure 2.8 Ternary-phase diagram of a-C:H deposited by LP-PECVD, showing material state as a function of sp^2, sp^3, and hydrogen concentration in the films. Reprinted from Ref. [26], Copyright (2002), with permission from Elsevier.

Carbon nitride films are also grown using gas mixtures of argon, methane, and nitrogen in PECVD systems. Due to their special properties (high hardness, low friction, chemical inertness, water resistivity, resistance to wear and corrosion, biocompatibility, and photo- and electroluminescence), such films have been the object of intensive research as a very promising way of surface modification [34].

Since the last decade, amorphous carbon film growth by PECVD at AP has also been intensively investigated. Although film hardness cannot reach typical values of LP processes because of ion thermalization, further applications have been envisaged of AP-PECVD amorphous carbon. Suzuki and Kodama reported the production of a-C:H films on large area substrates for gas barrier applications using line-type CVD equipment [35]. There, the plasma is sustained in the DBD configuration with a moving substrate working as the lower electrode. The deposition rate was around 1 μm/s, which is much larger than in the LP counterpart. The main concern lies in avoiding the formation of macroparticles that enhance roughness and degrade the mechanical properties of the films. In this approach, the carbon precursor C_2H_2 was diluted in Ar and N_2. Further studies have considered the addition of other gases, like H_2 and He [36]. Normally, such discharges are held at room temperature using excitation frequencies in the kHz–MHz range.

Carbon nanotubes (CNTs) are also well known for their high thermal and electrical conductivities and, particularly, for their excellent electron emission properties under high electric fields, which make them very attractive for flat panel displays and biomedical applications. CNTs were initially grown by an arc discharge showing extreme properties, like high mechanical resistance and tunable electric conductivity, within a wide interval. Nowadays, it is possible to generate "forests" of CNTs and carbon nanofibers (CNFs) with low-temperature plasmas using hydrocarbon precursors, which are decomposed into radicals that grow selectively on metal catalyst islands. The use of CVD techniques generally requires working at relatively high temperatures (>600°C), resulting in "spaghetti-like" films, in contrast to LP-PECVD, where the CNT and CNF axis is aligned by the electric field existing at the plasma sheath, thus generating the so-called vertically aligned nanofibers (Fig. 2.9) [37]. However, the application in flat panel displays requires deposition temperatures below the softening temperature of glass substrates (<550°C). Within this perspective, Kim et al. have successfully developed a DBD source of high charge density working under AP conditions at 400°C [38]. For this purpose, they used a specially designed dielectric, made of alumina perforated with capillary holes, to uniformly distribute the gases. The gas precursors are C_2H_2 diluted in He, with the addition of N_2 or NH_3, to eliminate amorphous carbon deposits on top of the CNTs. The same strategy has been adopted to grow other nanomaterials, such as vertically aligned boron nitride nanosheets. In this case, a gas mixture of $BF_3/N_2/H_2$ was used [39].

Finally, 2D nanomaterials, for instant flakes with a few layers of graphene, have been deposited by remote ECR-CVD with substrate temperatures substantially lower than in conventional CVD [40]. In this technique, the plasma species generated in the ECR chamber were transported through a turbulent flow toward the deposition chamber. There, two polarized electrodes discriminated neutral species to be deposited onto the substrate. The synthesis of low-temperature graphene films was achieved by means of H_2-aided fragmentation of C_2H_2 in a two-step process involving variations in pressure and substrate temperature. The crystalline domain size was between 50 nm and more than 300 nm, and the sheet resistance was lower than 2 kΩ sq^{-1}. However, these "low temperatures" are on the order of 500°C, still incompatible for temperature-sensitive substrates like polymers and organic tissues (biomedical applications). Therefore, finding recipes with lower temperature deposition of large, high-quality (monocrystalline) atomic monolayers onto any substrate is still an important issue in the synthesis of nanomaterials.

Figure 2.9 SEM micrograph of vertically aligned carbon nanofibers (CNFs) grown by PECVD. Reprinted from Ref. [37], with the permission of AIP Publishing.

2.5.2 Silicon-Based Compounds

Amorphous silicon (a-Si:H) films are grown using mixtures of SiH_4 and hydrogen to balance the density of dangling bonds within the atomic structure [41]. The use of additional precursors during deposition, or even a plasma post treatment, is carried out to dope the semiconductor for electronic applications. Another important issue is the maximization of quantum efficiency in the performance of a-Si:H in photovoltaic applications. Further research is also necessary to adapt PECVD parameters for the growth of a-Si:H on flexible substrates for the production of flexible solar cells. Another current trend is the coating of textured substrates for light-trapping purposes.

PECVD is being extensively used to deposit silicon-based films at LP. For instance, silicon nitride (Si_3N_4) is a dense material very adequate as a gas diffusion barrier and for protective applications. The films can be grown from a mixture of SiH_4 and NH_3 as reactive precursors. Organosilicon compounds like, tetramethylsilane, can also be used as an alternative to the toxic and inflammable SiH_4 precursor. a-SiN:H coatings with relatively good properties have been deposited on substrates from room temperature to ~100°C [42]. The use of two sources at RF and ultra-high-frequency in a CCP chamber at moderate pressures (0.5 mbar) provides better control of the plasma chemistry. This technique is being investigated to deposit a-SiN:H films with improved functional properties [43].

SiO_2 is another material coating well known for its gas barrier performance. Its application in the food packaging industry is a hot topic nowadays. Plasma sources based on the plasma-line concept have been developed to coat the interiors of beverage bottles and food packages. The plasma-line antenna consists basically of a quartz tube (used for gas feeding) with an embedded coaxial thin Cu tube. In particular, the antenna is inserted into a polyethylene terephthalate bottle to create an inner coating with a uniform SiO_2 thin layer. The microwave power is provided at the bottleneck and is transmitted inward through the antenna. This source acts as a transmission line, where the Cu core and the ignited external plasma play the respective roles of inner and outer conductors [44]. The source ignites a discharge inside the bottle with an atmosphere of hexamethyldisiloxane (HMDSO) and O_2 at a few tenths of mbars. The HMDSO/O_2 ratio must be minimized to obtain coatings with the desired stoichiometry and composition. The density of surface defects (pores) must be also minimal to avoid gas permeability.

An attractive alternative to LP SiO_2 deposition is the AP techniques, where no vacuum systems are required and batch processing could be included in a production line. Several approaches to obtaining SiO_2 at AP have been studied. As stated above, DBD is a well-established method to generate AP plasmas for the treatment of large surfaces [45]. These discharges are driven with voltages of some kilovolts at kHz frequency using two parallel plate electrodes covered with alumina to serve as a dielectric barrier. Usually, HMDSO or tetraethoxysilane (TEOS) is used as a gas precursor diluted in He, as well as N_2 as an inert diluted gas. Sometimes, argon and oxygen are also added to the source gases [46]. The carbon content in the SiO_2 thin films depends on the HMDSO or TEOS flow rate [47]. An additional N_2 plasma process after deposition can be necessary to remove the amorphous carbon incorporated in the films. It is also possible to produce dense SiO_2 thin films with a DBD source using $Ar/NH_3/SiH_4$ gas mixtures and RF or low-frequency glow discharges, although silane handling is highly hazardous [48].

2.5.3 Titanium-Based Compounds

Titanium dioxide (TiO_2) is an attractive material because of its physical and chemical properties, applicable in a wide area of technologies, such as solar cells, glass and ceramic self-cleaning, electronic component, and photocatalysis. The efficient use of these outstanding functional properties (namely, high refractive index, optical transmittance in the visible and infrared range, nontoxicity, biocompatibility, etc.) relies strongly on the material's structural and morphological characteristics, which are affected by the deposition parameters, particularly temperature. This fact demands growth techniques with a large choice of substrate temperatures, such as LP-PECVD [49]. Typical Ti-containing precursors are $TiCl_4$ and Ti alkoxides, although precursors with more complex structures have also been tested for the sake of improved functionality in catalysis applications. Atmospheric plasma techniques, such as the DBD plasma jet [50] or RF atmospheric plasma torch [51], have also been applied for the deposition of TiO_2 films. A drawback of AP plasmas is that the mean free path of reactive species is very short (a few micrometers) compared to that of LP plasmas (several millimeters), which makes it difficult to control the final properties of the grown films. However, in recent years, DBD has been proved to be an accurate method for TiO_2 synthesis for photocatalytic applications, using as gas precursors Ti chemical compounds very similar to those employed at LP (usually titanium tetraisopropoxide and $TiCl_4$) [52].

On the other hand, owing to their excellent properties (high hardness and melting temperature, good resistance to corrosion and wear, low electrical and thermal resistance, high thermal stability, etc.), TiN and TiC are widely applied on cutting tools, diffusion barriers, electrical contacts, and decorative coatings [53]. In most of these applications, the coatings are deposited by the well-known sputtering techniques. In mass production applications, the vacuum chamber is provided with a rotatory support of the samples to get a homogeneous coating thickness, generally on small, 3D substrates. However, the application of these techniques for forming tools (molds and dies) poses very serious limitations to the above-mentioned procedures due to their complex shapes as well as the heavy weight of these tools. In this case, LP-PECVD discharges working at about 500°C, in the range of 10–100 Pa, are a suitable method for depositing hard Ti compounds (TiN, TiC, TiCN, etc.) with very good thickness homogeneity, even on those sites out of line of sight of the gas flow. Typical reaction gases are $TiCl_4$, CH_4, N_2, H_2, and Ar, and the temperature is about 500°C [54, 55].

TiN thin films can be also prepared by plasma-enhanced atomic layer deposition from tetrakis (dimethylamido) titanium. In this case, the reactive gas plays an important role: nitrogen-based plasma (N_2 and NH_3) results in low oxygen (~3%) and carbon (~2%) contaminations and a well-defined columnar grain structure. An excess of N_2 (~4%) was found in the films deposited using N_2 plasmas. Deposition with H_2 plasma resulted in higher carbon and oxygen contaminations (~6% for each element) [56].

Regarding optoelectronic applications, TiN films are widely investigated due to their low resistivity and high light absorption. However, an optimal combination with transparent TiO_2 can be used to obtain low emissive coatings for solar-energy saving in glass panels of architectural windows [57]. Particularly, the atmospheric plasma deposition process enables the deposition of TiN-TiO_2 hybrid films on large and/or complex shaped substrates [58]. The precursor can consist of titanium ethoxide (or another titanium metal–organic compound), usually with He as the carrier gas.

References

1. S. E. Alexandrov, M. Hichman, Chemical vapour deposition enhanced by atmospheric pressure non-thermal non-equilibrium plasmas, *Chem. Vap. Depos.*, **11** (2005) 457–468.

2. L. Martinu, O. Zabeida, J. E. Klemberg-Sapieha, Plasma-enhanced chemical vapor deposition of functional coatings, in *Handbook of*

Deposition Technologies for Films and Coatings, 3rd ed., ed. P. M. Martin, Elsevier, Amsterdam (2010), p. 392.

3. D. Merche, N. Vandencasteele, F. Reniers, Atmospheric plasmas for thin film deposition: a critical review, *Thin Solid Films*, **520** (2012) 4219–4236.
4. L. Bárdos, H. Baránková, Cold atmospheric plasma: sources, processes, and applications, *Thin Solid Films*, **518** (2010) 6705–6713 .
5. F. Massines, C. Sarra-Bournet, F. Fanelli, Atmospheric pressure low temperature direct plasma technology: status and challenges for thin film deposition, *Plasma Processes Polym.*, **9** (2012) 1041–1073.
6. C. Tendero, C. Tixier, P. Tristant, J. Desmaison, P. Leprince, Atmospheric pressure plasmas: a review, *Spectrochim. Acta, Part B*, **61** (2006) 2–30.
7. R. A. Baragiola, P. Riccardi, Electron emission from surfaces induced by slow ions and atoms, in *Reactive Sputter Deposition*, eds. D. Depla, S. Mahieu, Springer, Berlin (2008), pp. 43–60.
8. M. Keidar, I. I. Beilis, in *Plasma Engineering*, 2nd ed., Elsevier, London (2018).
9. C. Corbella, M Rubio-Roy, E. Bertran, S. Portal, E. Pascual, M. C. Polo, J. L. Andujar, Ion energy distributions in bipolar pulsed-dc discharges of methane measured at the biased cathode, *Plasma Sources Sci. Technol.*, **20** (2011) 015006.
10. H. Kakiuchi, H. Ohmi, K. Yasukate, Atmospheric-pressure low-temperature plasma processes for thin film deposition, *J. Vac. Sci. Technol., A*, **32** (2014) 030801.
11. S. Kechkart, Experimental investigation of a low pressure capacitively-coupled discharge, Doctoral thesis, Dublin City University (2015).
12. A. Bogaerts, E. Neyts, R. Gijbels, J. van der Mullen, Gas discharge plasmas and their applications, *Spectrochim. Acta, Part B*, **57** (2002) 609–665.
13. M. A. Lieberman, A. J. Lichtenberg, *Principles of Plasma Discharges and Materials Processing*, Wiley Interscience (2005).
14. H. Conrads, M. Schmidt, Plasma generation and plasma sources, *Plasma Sources Sci. Technol.*, **9** (2000) 441–454
15. U. Helmersson, M. Lattemann, J. Bohlmark, A. P. Ehiasarian, J. T. Gudmundsson, Ionized physical vapor deposition (IPVD): a review of technology and applications, *Thin Solid Films*, **513** (2006) 1–24.
16. J. L. Andujar, M. Vives, C. Corbella, E. Bertran, Growth of hydrogenated amorphous carbon films in pulsed d.c. methane discharges, *Diamond Relat. Mater.*, **12** (2003) 98–104.

17. A. Aijaz, S, Louring, D, Lundin, T. Kubart, J. Jensen, K. Sarakinos, U. Helmersson, Synthesis of hydrogenated diamond-like carbon thin films using neon-acetylene based high power impulse magnetron sputtering discharges, *J. Vac. Sci. Technol., A*, **34** (2016) 061504.
18. C. Corbella, I. Bialuch, M. Kleinschmidt, K. Bewilogua, Modified DLC coatings prepared in a large-scale reactor by dual microwave/pulsed-DC plasma-activated chemical vapour deposition, *Thin Solid Films*, **517** (2008) 1125–1130.
19. Z. Chen, S. Rauf, K. Collins, Self-consistent electrodynamics of large-area high-frequency capacitive plasma discharge, *J. Appl. Phys.*, **108** (2010) 073301.
20. D. Eremin, S Bienholz, D. Szeremley, J. Trieschmann, S. Ries, P. Awakowicz, T. Mussenbrock, R. P. Brinkmann, On the physics of a large CCP discharge, *Plasma Sources Sci. Technol.*, **25** (2016) 025020.
21. B. G. Heil, U, Czarnetzki1, R. P. Brinkmann, T. Mussenbrock, On the possibility of making a geometrically symmetric RF-CCP discharge electrically asymmetric, *J. Phys. D: Appl. Phys.*, **41** (2008) 165202.
22. D. Hegemann, B. Nisol, S. Watson, M. R. Wertheimer, Energy conversion efficiency in plasma polymerization: a comparison of low- and atmospheric-pressure processes, *Plasma Processes Polym.*, **13** (2016) 834–842.
23. R. Brandenburg, Dielectric barrier discharges: progress on plasma sources and on the understanding of regimes and single filaments, *Plasma Sources Sci. Technol.*, **26** (2017) 053001.
24. D. Pappas, Status and potential of atmospheric plasma processing of materials, *J. Vac. Sci. Technol., A*, **29** (2011) 208011– 208017.
25. J. Salge, Plasma-assisted deposition at atmospheric pressure, *Surf. Coat. Technol.*, **80** (1996) 1–7.
26. J. Robertson, Diamond-like amorphous carbon, *Mater. Sci. Eng., R*, **37** (2002) 129–282.
27. K. Bewilogua, D. Hofmann, History of diamond-like carbon films: from first experiments to worldwide applications, *Surf. Coat. Technol.*, **242** (2014) 214–225.
28. R. K. Roy, K.-R. Lee, Biomedical applications of diamond-like carbon coatings: a review, *J. Biomed. Mater. Res. Part B*, **83** (2007) 72–84.
29. V. Singh, J. C. Jianga, E. I. Meletis, Cr-diamondlike carbon nanocomposite films: synthesis, characterization and properties, *Thin Solid Films*, **489** (2005) 150–158.

30. C. Corbella, E. Bertran, M. C. Polo, E. Pascual, J. L. Andújar, Structural effects of nanocomposite films of amorphous carbon and metal deposited by pulsed-DC reactive magnetron sputtering, *Diamond Relat. Mater.*, **16** (2007) 1828–1834.

31. A. Pardo, Metal incorporation in hydrogenated amorphous carbon films deposited by biased ECR-CVD, Doctoral thesis, Universidad Autónoma de Madrid (2012).

32. J. G. Buijnsters, M. Camero, L. Vázquez, F. Agulló-Rueda, R. Gago, I. Jiménez, C. Gómez-Aleixandre, J. M. Albella, Tribological study of hydrogenated amorphous carbon films with tailored microstructure and composition produced by bias-enhanced plasma chemical vapor deposition, *Diamond Relat. Mater.*, **19** (2010) 1093–1102.

33. D. Martinez-Martinez, J. Th. M. De Hosson, On the deposition and properties of DLC protective coatings on elastomers: a critical review, *Surf. Coat. Technol.*, **258** (2014) 677–690.

34. M. Camero, R. Gago, C. Gómez-Aleixandre, J. M. Albella, Hydrogen incorporation in CNx films deposited by ECR chemical vapour deposition, *Diamond Relat. Mater.*, **12** (2003) 632–635.

35. T. Suzuki, H. Kodama, Diamond-like carbon films synthesized under atmospheric pressure synthesized on PET substrates, *Diamond Relat. Mater.*, **18** (2009) 990–994.

36. A. M. Ladwig, R. D. Koch, E. G. Wenski, R. F. Hicks, Atmospheric plasma deposition of diamond-like carbon coatings, *Diamond Relat. Mater.*, **18** (2009) 1129–1133.

37. V. I. Merkulov, A. V. Melechko, M. A. Guillom, D. H. Lowndes, Effects of spatial separation on the growth of vertically aligned carbon nanofibers produced by plasma-enhanced chemical vapor deposition, *Appl. Phys. Lett.*, **30** (2002) 476–478.

38. K. N. Kim, S. M. Lee, A. Mishra, G. Y. Yeom, Atmospheric pressure plasmas for surface modification of flexible and printed electronic devices: a review, *Thin Solid Films*, **598** (2016) 315–334.

39. J. Yu, L. Qin, Y. Hao. S. Kuang, X. Bai, Y.-M. Chong, W. Zhang, E. Wang, Vertically aligned boron nitride nanosheets: chemical vapor synthesis, ultraviolet light emission, and superhydrophobicity, *ACS Nano*, **4** (2010) 414–422.

40. R. Muñoz, L. Martínez, E. López-Elvira, C. Munuera, Y. Huttel, M. García-Hernández, Direct synthesis of graphene on silicon oxide by low temperature plasma enhanced chemical vapour deposition, *Nanoscale*, **10** (2018) 12779.

41. J. Ramanujam, A. Verma, Photovoltaic properties of a-Si:H films grown by plasma enhanced chemical vapor deposition: a review, *Mater. Express*, **2** (2012) 177–196.

42. C. Yang. J. Pham, Characteristic study of silicon nitride films deposited by LPCVD and PECVD, *Silicon*, **10** (2018) 2561–2567.

43. B. B. Sahu, J. G. Han, Electron heating mode transition induced by mixing radio frequency and ultrahigh frequency dual frequency powers in capacitive discharges, *Phys. Plasma*, **23** (2016) 053514.

44. M. Deilmann, H. Halfmann, S. Steves, N. Bibinov, P. Awakowicz, Silicon oxide permeation barrier coating and plasma sterilization of PET bottles and foils, *Plasma Processes Polym.*, **6** (2009) S695–S699.

45. R. Reuter, N. Gherardi, and J. Benedikt, Effect of N_2 dielectric barrier discharge treatment on the composition of verythin SiO_2-like films deposited from hexamethyl-disiloxane at atmospheric pressure, *Appl. Phys. Lett.*, **101** (2012) 194104.

46. R. Wang, W. Li, Ch. Zhang, Ch. Ren, Kostya (Ken) Ostrikov, T. Shao, Thin insulating film deposition on copper by atmospheric-pressure plasmas, *Plasma Processes Polym.*, **14** (2017) 1770011.

47. V. Raballand, J. Benedikt, and A. von Keudell, Deposition of carbon-free silicon dioxide from pure hexamethyldisiloxane using an atmospheric microplasma jet, *Appl. Phys. Lett.*, **92** (2008) 091502.

48. R. Bazinette, J. Paillol, J. F. Lelièvre, F. Massines, Atmospheric pressure radio-frequency DBD deposition of dense silicon dioxide thin film, *Plasma Processes Polym.*, **13** (2016) 1015.

49. N. C. da Cruz, E. C. Rangel, J. Wang, B. C. Trasferetti, M. A. B. de Moraes, Properties of titanium oxide films obtained by PECVD, *Surf. Coat. Technol.*, **126** (2000) 123–130.

50. H.-K. Seo, C. M. Elliott, H.-S. Shin, Mesoporous TiO_2 films fabricated using atmospheric pressure dielectric barrier discharge jet, *ACS Appl. Mater. Interfaces*, **2** (2010) 3397.

51. S. Collette, J. Hubert, A. Batan, K. Baert, M. Raes, I. Vandendael, A. Daniel, C. Archambeau, H. Terryn, F. Reniers, Photocatalytic TiO_2 thin films synthesized by the post-discharge of an RF atmospheric plasma torch, *Surf. Coat. Technol.*, **289** (2016) 172–178.

52. Q. Chen, Q. Liu, A. Ozkan, B. Chattopadhyay, G. Wallaert, K. Baert, H. Terryn, M. P. Delplancke-Ogletree, Y. Geerts, F. Reniers, Atmospheric pressure dielectric barrier discharge synthesis of morphology-controllable TiO_2 films with enhanced photocatalytic activity, *Thin Solid Films*, **664** (2018) 90–99.

53. X. Lin, G. Zhao, L. Wu, G. Duan, G. Han, TiN_x thin films for energy-saving application prepared by atmospheric pressure chemical vapor deposition, *J. Alloys Compd.*, **502** (2010) 195–198.

54. C. Mitterer, F. Holler, F. Ustel, D. Heim, Application of hard coatings in aluminium die casting—soldering, erosion and thermal fatigue behaviour, *Surf. Coat. Technol.*, **125** (2000) 233–239.

55. C. Mitterer, F. Holler, Reitberger, E. Badisch, M. Stoiber, C. Lugmair, R. Nöbauer, Th. Müller, R. Kullmer, Industrial applications of PACVD hard coatings, *Surf. Coat. Technol.*, **163–64** (2003) 716–722.

56. I. Krylov, X. Xu, E. Zoubenko, K. Weinfeld, S. Boyeras, F. Palumbo, M. Eizenberg, D. Ritte, Role of reactive gas on the structure and properties of titanium nitride films grown by plasma enhanced atomic layer deposition, *J. Vac. Sci. Technol., A*, **36** (2018) 06A105.

57. X. Lin, G. Zhao, L. Wu, G. Duan, G. Han, TiN_x thin films for energy-saving application prepared by atmospheric pressure chemical vapor deposition, *J. Alloys Compd.*, **502** (2010) 195–198.

58. S. Dong, M. Watanabe, R. H. Dauskard, Conductive transparent TiN_x/TiO_2 hybrid films deposited on plastics in air using atmospheric plasma processing, *Adv. Funct. Mater.*, **24** (2014) 3075–3081.

Chapter 3

Deposition of Porous Nanocolumnar Thin Films by Magnetron Sputtering

R. Alvarez,[a,b] A. R. González-Elipe,[a] and A. Palmero[a]
[a]*Nanotechnology on Surfaces and Plasma Laboratory, Instituto de Ciencia de Materiales de Sevilla (CSIC-Univ. Sevilla), Avda. Américo Vespucio 49, 41092 Sevilla, Spain*
[b]*Departamento de Física Aplicada I, Escuela Politécnica Superior, Universidad de Sevilla, c/ Virgen de África 7, 41011 Seville, Spain*
arge@icmse.cisc.es; alberto.palmero@csic.es; rafael.alvarez@icmse.csic.es

3.1 Introduction to Magnetron Sputtering

Physical vapor deposition (PVD) encompasses a large variety of thin-film growth methods involving the ejection of atoms from a target and their arrival to a substrate, where these species accumulate in the form of a film. A typical PVD procedure consists of the evaporation from a heated target (thermally or by electron beam bombardment), while in the industry, the most common approach relies on the sputtering of atoms from a solid target, typically mediated by the action of some permanent magnets (i.e., magnetron sputtering [MS]). In a typical MS process, a metal target

Plasma Applications for Material Modification: From Microelectronics to Biological Materials
Edited by Francisco L. Tabarés

ISBN 978-981-4877-35-0 (Hardcover), 978-1-003-11920-3 (eBook)
www.jennystanford.com

immersed in Ar or another noble gas at a low pressure is negatively polarized to several hundred volts to induce the partial ionization of the gas and the formation of a plasma. The positively ionized Ar^+ ions in the plasma are accelerated toward the negatively polarized target, where they induce the sputtering of the target atoms and their deposition on a substrate located in front of the target [1].

Unlike the low energy carried by the atoms evaporated from a target (the typical kinetic energy is in the order of 0.1 eV), sputtered atoms may possess energies of several electron volts, a factor that should contribute to densify the deposited films. Densification and compactness are in fact the expected outputs of the MS procedure, features that are enhanced in the modern High Power Impulse Magnetron Sputtering [2], a variation of this technique A large variety of experimental results sustain the application of sputtering techniques for the preparation of compact thin films of interest as protective layers, passive optical applications or hard coatings. These first considerations might suggest that MS techniques cannot be used for the preparation of porous and nanostructured thin films. However, the objective of this chapter is precisely to prove that MS, utilized under certain conditions, is suitable for the preparation of these kinds of low-density thin films. To give credit to this possibility, in this chapter, we will discuss the following issues that define its content index:

- Nanostructuring variables during thin-film sputtering deposition
- The sputtering mechanism and properties of the sputtered atoms
- Transport of the sputtered atoms through the plasma gas
- Reactive magnetron sputtering (r-MS)
- Deposition geometry as a variable to induce the growth of nanocolumnar thin films
- Process control and description of thin-film growth
- Case studies: Application of porous and nanostructured thin films prepared by MS

Before proceeding to a specific analysis of these topics, let us look at some basic issues of the MS techniques that are of relevance for understating these points. A complete description of these questions can be found in relevant books and reviews on this subject [1, 3].

In practice, the sputtering process is in most cases enhanced by the application of magnets onto the back of the target. The induced magnetic field contributes to focussing the plasma electrons around the target, thus increasing their lifetime and leading to the enhancement of the Ar ionization rate. Typical operation voltages in MS are lower than in normal sputtering without magnets and the sputtering efficiency larger. MS targets may adopt different shapes and sizes. For example, they can be planar with rectangular or circular shapes or cylindrical and can be fixed or rotate during operation. In laboratory applications, the targets are circular and may vary from one to several inches in size. In the industry, the rectangular shape is more common and the size may reach up to several meters. An important consequence of the application of magnets on the back of the target is to concentrate the ion bombardment on a certain region of the target, the so-called racetrack, while the rest of the target remains almost unaffected by the impingement of ions (for more details about the effect of magnets and the different forms to apply them, see Ref. [3]). In a circular target, the racetrack adopts the form of a ring. A scheme of a typical MS geometry, specifically considering the racetrack in the target is displayed in Fig. 3.1. The geometrical considerations that are the basis of important thin-film nanostructuration principles refer to the dimensions of this racetrack and the orientation and distance of the substrate with regard to the actual location and orientation of the former (see Sections 3.2.1 and 3.2.2).

3.1.1 Nanostructuring Variables during Thin-Film Sputtering Deposition

In 1974, Thornton [4] proposed the classical zone model to account for the evolution of the microstructure of magnetron-sputtered thin films as a function of two key variables of the deposition process, the temperature of the substrate and the pressure of the plasma gas. Basic tendencies predicted by the zone model are an increase in densification and domain size with the substrate temperature and a tendency toward little nanocolumns and higher porosity as the pressure of the plasma gas increases.

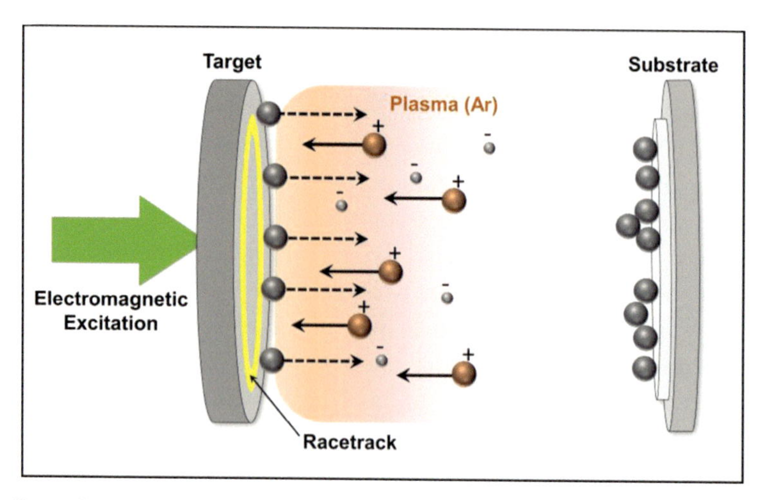

Figure 3.1 Typical MS geometry using a circular target and description of the sputtering process, with indication of the racetrack made by the ions impinging on the surface of the target. Ar^+ ions from the plasma impinge on the target, leading to the sputtering of the atoms that eventually arrive and accumulate on the substrate.

A key feature for the nanostructuration of thin films that will be the basis of the present chapter is the change in the orientation of the substrate plane with respect to the target. In a conventional deposition procedure, as that described by Thornton [4], the substrate is placed parallel to the target surface and a significant fraction of the sputtered particles arrives at the substrate in a normal or close to normal direction (c.f., Fig. 3.1). A key issue for the structuration of magnetron sputtered thin films in the form of nanocolumns is to modify the deposition geometry, placing the substrate at an oblique angle with respect to the racetrack of the target. Under this configuration, the sputtered particles arrive at the substrate at a glancing angle and give rise to a particular growth regime that, in most cases, is characterized by the formation of separated tilted nanocolumns [5]. This way of depositing a thin film is usually denoted as "oblique angle deposition" (OAD) or, sometimes "glancing angle deposition". The second term was coined to describe the deposition of evaporated thin films, a process in which the evaporated particles move in vacuum from the target to the substrate and, therefore, do not interact with gas molecules, as it may happen in the sputtering techniques. The effect on the film growth of this substantial difference, as well as the aforementioned difference in energy of evaporated or sputtered particles, will be the object of a thorough discussion in Sections 3.2.1 and 3.2.5.

Unlike the evaporation techniques, where the chemical state of the deposited film generally coincides with that of the evaporation target (e.g., metal, semiconductor, and, less straightforward, oxide), the chemical state of sputtered thin films can be altered by reaction with the plasma gas. For example, oxides or nitrides can be prepared using metal targets if oxygen or nitrogen is added to the plasma gas, typically Ar. This process is generally known as reactive magnetron sputtering (r-MS), a process that involves a series of experimental questions, such as target poisoning, changes in sputtering rate or the formation of intermediate low stoichiometry phases, all of which must be considered to fully account for the deposition process. The interplay in an r-MS experiment between the oblique angle geometry of deposition, the concentration of the reactive gas, or other experimental variables with the nanostructuration mechanisms of the thin films will be also discussed in detail in this chapter (Section 3.2.5).

Implicitly, the said nanostructuration variables in OAD apply to the deposition of thin films onto flat substrates. That additional nanostructuration processes may intervene during the deposition and growth of thin films for patterned nonflat substrates will be revised in Section 3.2.6. In this case the arrival angle of the sputtered particles changes at a local scale, giving rise to additional nanostructuration processes and, hence, new thin-film morphologies.

3.1.2 The Sputtering Mechanism

Among all the fundamental processes influencing film deposition and the formation of the film nanostructure, the plasma-assisted sputtering of target species emerges as a key relevant process, defining not only the kinetic energy of the deposition species but also their angular distribution. In general terms, plasma-assisted sputtering of a material can be analyzed within the same framework as the one that describes the ion irradiation of surfaces, which is a well-developed and mature topic. However, typical operational ranges in MS depositions restrain the application of this theory to conditions where the ion kinetic energy is below 1 KeV and the incidence rather perpendicular, which is briefly described next.

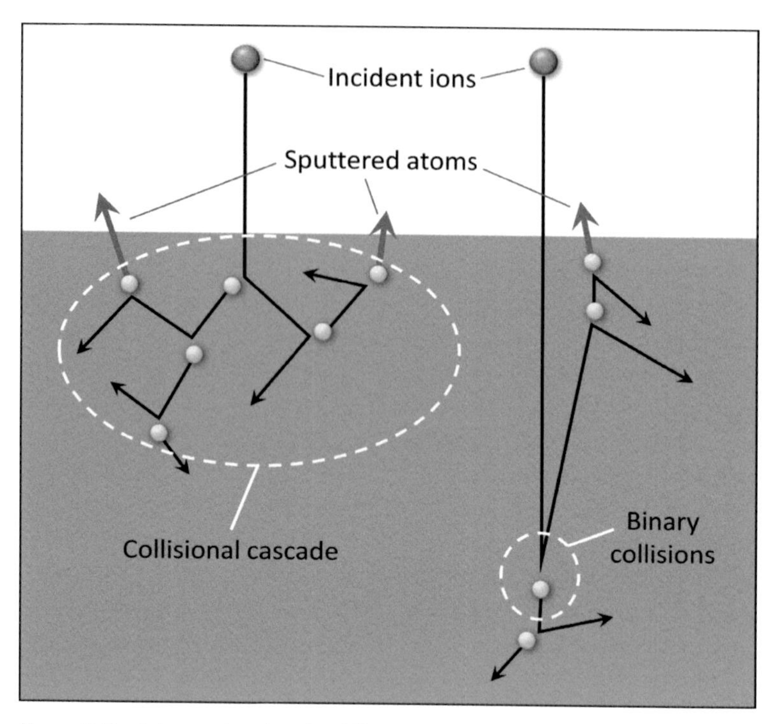

Figure 3.2 Scheme showing the different sputtering mechanisms from a solid target upon ion irradiation. On the left, the typical mechanism when the mass of the ion is much higher than that of the material is depicted: the incoming ions cause a full collisional cascade in the material's network, causing the ejection of some atoms in the first monolayers. On the right, the typical sputtering mechanism when the ions are much lighter than the material's atoms is depicted: ions rebound within the material via binary collisions, causing the ejection of species from the first monolayers.

From a physical point of view there are two basic mechanisms responsible for the sputtering of a material from a solid surface upon ion irradiation, depending on the relative masses of target and ion species (see Fig. 3.2): (i) when the plasma ion is much heavier, it transfers its energy to atoms in the first monolayers of the material, causing a collisional cascade that results in the ejection of some of them, and (ii) when the ion is much lighter, the collisional cascade can't be fully formed, and it penetrates the material a certain distance and gets reflected toward the surface, causing the ejection of some atoms after a collision. In either case, a key quantity defining the process is the so-called sputtering yield, $Y(E_i)$, accounting for the average number of sputtered atoms from the target per incident ion with kinetic energy, E_i. In general, $Y(E_i) = 0$ when E_i is below a certain threshold (20–30 eV), rapidly increasing its value for increasing energies (it reaches a maximum for energies much above the typical operational range of MS depositions) [6].

In addition to $Y(E_i)$, two key quantities define the MS deposition of thin films: the kinetic energy distribution of sputtered species, E_S, and their angular distribution of ejection, defined by a solid angle Ω. For this, a differential sputtering yield is introduced $Y(E_i, E_S, \Omega)dE_S d\Omega$ that accounts for both distributions. In this way, when the sputtering mechanism is associated with the existence of a collisional cascade, the so-called Thompson formula describes adequately the energy distribution:

$$Y(E_S)dE_S \propto \frac{E_S}{(E_S + U_S)^3} dE_S,$$

where U_S is the binding energy of surface atoms in the material. This formula states that the incident ion energy does not influence the energy distribution of sputtered species, which only depends on U_S. In fact, it peaks at a value of $E_S = U_S/2$, which points to a typical energy of sputtered species in the order of a few electron volts. Therefore, and according to the Thompson formula, the ion incident energy determines the average amount of sputtered species from the target but not their energy distribution. Moreover, the angular distribution of sputtered species from a well-developed collisional cascade is a typical cosine-type distribution, although an under-cosine or heart-shaped distribution can be found when the cascade does not fully develop. As we will see later, these two functions, the energy and angular distributions of sputtered atoms, are key to understanding the development of the film nanostructure and porosity.

A relevant aspect concerning the angular distribution of sputtered species relates to the state of the target and, in particular, to the racetrack profile. As mentioned in Section 3.1.1, this region of the target receives a preferential ion impingement and experiences intense sputtering, which in time might cause it to wear away, forming a valley on the target surface. This may indeed affect the angular distribution of sputtered species leaving the target, and could make the deposition process rather dependent on the aging of the target [6].

3.1.3 Transport of Sputtered Species in the Plasma: Thermalization Degree

As we have seen in the previous section, ion-induced collisions in the first monolayers of a target material cause the ejection of atoms with kinetic energies of a few electron volts, preferentially in a direction perpendicular to the target. However, before their deposition, these atoms must pass through the plasma, where they may experience various scattering processes with heavy species, which could affect their energy and angular distribution when arriving at the substrate. At this point, it is important to underline that typical magnetron plasmas employed to grow thin films possess an ionization degree in the order of 0.1%–0.01%, implying that most collisions of sputtered atoms with plasma species would involve neutral species, with a negligible amount of collisions with plasma ions [7]. Moreover, collisions with electrons are also negligible, not only because of their lower number in comparison with neutral species, but also because of their low mass, which is light enough to not significantly alter the kinetic energy and momentum of target atoms. Thus, when the collisional transport of sputtered species is being solely addressed, the plasma can be taken as a rarified gas made of neutral species, with density, n, calculated by the ideal gas law, $n = p_g/k_B T_g$, where p_g is the pressure, k_B is the Boltzmann constant, and T_g is the temperature. Moreover, and due to the nonthermal plasma nature of the magnetron plasma, these neutral species stay at a local thermodynamic equilibrium and at temperatures, T_g, of 300–400 K.

After being sputtered with typical energies of several electron volts, target atoms experience diverse scattering events with gas species, whose typical kinetic energy is in the thermal range (below 0.1 eV). These collisions, therefore, would progressively reduce the energy of the sputtered species and tend to randomize their angular distribution. Attending to their collisional transport, three types of deposition species can be distinguished:

- Ballistic species: These do not experience any collision on their way from the target to the substrate and arrive at the film surface with their original energy and along the angle aligning the racetrack and the film.
- Thermalized species: These experience a very high number of collisions before being deposited, becoming thermalized with

the gas, that is, their typical kinetic energy is in the thermal range (below 0.1 eV) and their momentum distribution is isotropic.

- Partially thermalized: These have experienced some collisions but not enough to reach full thermalization.

On the basis of these definitions, a new quantity, ν, can be introduced as the average number of collisions a sputtered atom needs to experience in the gas in order to become thermalized, a quantity that would depend on the relative masses between species, kinetic energies, atomic radii, etc. [8]. Moreover, the typical number of collisions a sputtered species experiences before being deposited can be determined using its mean free path, λ, which, according to the kinetic theory of gases, relates to the gas density through the formula $\lambda = 1/n\sigma$, where σ is the cross section for a collision, usually taken between hard spheres. In this way, if L is the distance between the target and the film, the quantity $L/\lambda\nu$ indicates the ratio between the typical number of collisions a sputtered species experiences before deposition, L/λ, and the average number of collisions that a sputtered species requires in order to become thermalized, ν. Due to its importance when defining the collisional transport, this ratio is called the thermalization degree of deposition species, $\Xi = L/\lambda\nu$. In this way, when $\Xi << 1$ sputtered atoms hardly experience any collision with gas species and most of them arrive at the substrate with their original kinetic energy and directionality, that is, most deposition species are ballistic. When $\Xi >> 1$, on the other hand, sputtered atoms undergo a high number of collisions and the deposition is carried out by thermalized species. Intermediate values of Ξ would indicate a different proportion of thermalized, ballistic, and near-thermalized species [9].

From a qualitative point of view, depositions in ballistic and thermalized regimens are very different. Ballistic species follow the direction aligning the racetrack and the film surface maintaining most of their original kinetic energy when sputtered, while thermalized species possess much lower kinetic energy and follow a different angular distribution function. In a thermalized regimen, this distribution function can be theoretically calculated knowing that the velocity distribution in the gas phase is isotropic, that is, $F(v, \theta, \varphi) \equiv F(v)$, where v is the modulus of the velocity and θ and φ

the polar and azimuthal angles, respectively, defined with respect to the substrate in Fig. 3.3. Hence, the angular distribution of arrival, $f(\theta, \varphi)$, is calculated as

$$f(\theta,\varphi)\mathrm{d}\theta\mathrm{d}\varphi = \sin\theta\mathrm{d}\theta\mathrm{d}\varphi\int \mathrm{d}v v^2 F(v)(\vec{v}\cdot\vec{s}) \propto \sin\theta\cos\theta\mathrm{d}\theta\mathrm{d}\varphi,$$

where $\vec{s}$ is a unitary vector normal to the substrate. Typical angular distributions of ballistic and thermalized species as a function of θ are shown in Fig. 3.4. There, the ballistic peak is centered on the angle aligning the racetrack and the substrate, which we have assumed to be ~17°, and possesses a specific width due to the existence of an inner and outer radii defining the racetrack. The distribution corresponding to the thermalized species is broader, covering all the angles from 0° to 90°, presenting a maximum at an incident angle of 45°. In intermediate conditions, both contributions will be present as expected for the arrival of species in the partially thermalized state.

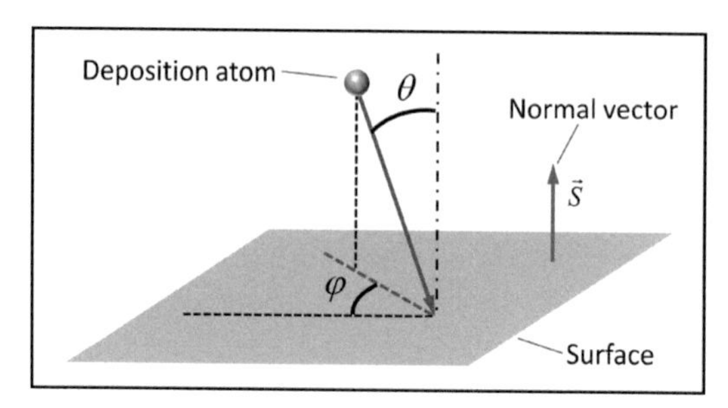

Figure 3.3 Definition of the polar (θ) and azimuthal (φ) angles with respect to a surface. $\vec{s}$ is a vector perpendicular to the surface and modulus 1.

The concept of thermalization degree introduces a very useful quantity to qualitatively assess the energy and angular distribution of the species upon deposition. In terms of more fundamental quantities, it can be calculated as

$$\Xi = \frac{L}{\lambda v} = \frac{n\sigma L}{v} = \frac{\sigma}{k_B T_g v} p_g L = K p_g L,$$

where $K = \sigma / k_B T_g v$ is a quantity dependent on the conditions of the gas (through T_g), the nature of the species involved (through σ and v), and the initial kinetic energy of the sputtered species (through v). In this way, for gas and a target material of a given nature, the higher the pressure or the distance between the target and the substrate, the more collisions the sputtered species would experience, the lower the kinetic energy they would have, and the more isotropic the angular distribution function would be in the plasma gas. On the basis of this concept, there is a straightforward relation between the deposition rate in classical MS configuration, r, and the thermalization degree by means of the so-called Keller–Simmons formula, which originally took the form $r = \Phi_0 \frac{p_0 L_0}{p_g L}\left[1 - \exp\left(-\frac{p_g L}{p_0 L_0}\right)\right]$, where, Φ_0 is the number of sputtered species per unit time and surface area and $p_0 L_0$ the so-called throw distance, an adjustable parameter that depends on the deposition conditions [10]. Interestingly, when the gas temperature could be considered constant throughout the discharge, it was demonstrated the connection $\Xi = p_g L / p_0 L_0$, finding the relation $r = \Phi_0 \frac{1 - \exp(-\Xi)}{\Xi}$.

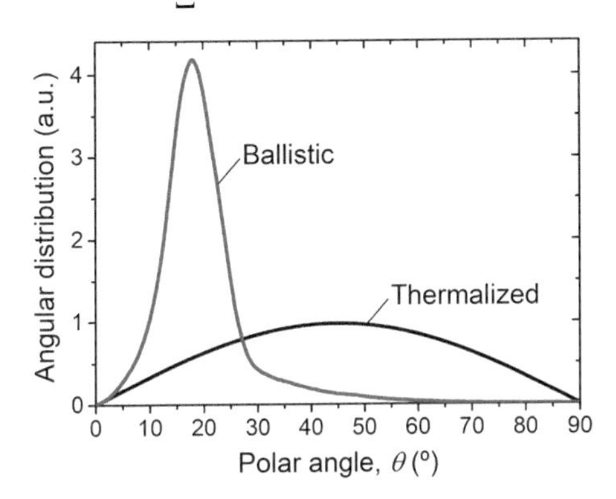

Figure 3.4 Typical distribution of ballistic and thermalized sputtered species. In the former case, atoms arrive at the surface along the direction aligning the substrate and the racetrack. In the latter, species move following a Brownian-like behavior, with an isotropic momentum distribution in the gas phase.

3.1.4 Reactive Magnetron Sputtering

On occasion, the chemical composition of the coatings is required to be different from that of the target, for instance, when the latter is purely metallic and the film requires to grow containing certain amount of oxygen or nitrogen. In this case, a small amount of reactive gas can be added to the plasma to alter the film composition during growth. This variation of the technique is commonly known as r-MS, and it is widely employed for the production of oxides, nitrides, and other compound thin films. However, and despite its simplicity, the practical application of this procedure poses some serious drawbacks related to the existence of low deposition rates when growing fully stoichiometric layers that are associated with an undesired phenomenon called "target poisoning" [6].

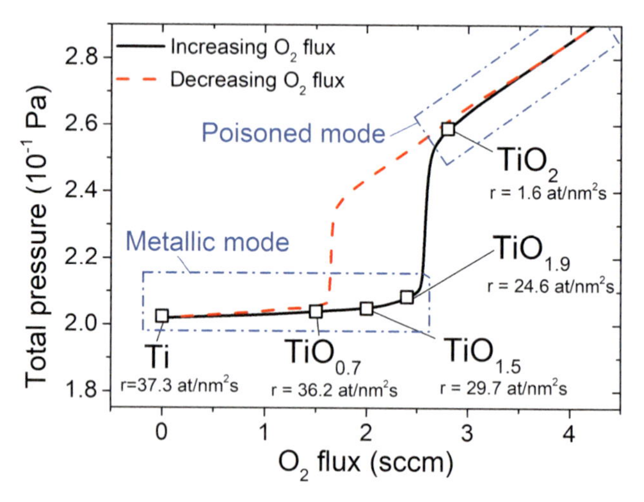

Figure 3.5 Typical hysteresis curve of the total pressure in the deposition reactor as a function of the flux of the reactive gas in the reactive magnetron sputtering deposition of TiO_x thin films. In the metallic mode, the target remains clean and, hence, the deposition rate is high and the film stoichiometry increases with the amount of oxygen in the reactor. In the poisoned mode, the target becomes poisoned when the film stoichiometry is almost saturated with oxygen. Under this condition, the deposition rate drops to much lower values than in the metallic mode.

Target poisoning accounts for changes in the chemical composition of the first monolayers of the target material due to the presence of a reactive gas in the plasma. This phenomenon is responsible for a well-known hysteresis behavior of the pressure in the reactor when solely increasing/decreasing the flux of reactive gas in the plasma. In Fig. 3.5 we show a case example of deposition of TiO_x thin films (using argon as a nonreactive gas, oxygen as a reactive gas and a titanium target), where we depict the changes in the total pressure in the reactor as a function of the oxygen flux, along with the oxygen content in the film (film composition) and the deposition rate at some particular conditions.

As depicted in this figure, different regions can be distinguished for increasing oxygen fluxes and the oxidation state of the target:

- Metallic mode (below 2.5 sccm in Fig. 3.5): In this region, the total pressure in the reactor changes slightly with the oxygen flux, mainly due to the existence of a much larger amount of argon and because most oxygen in the discharge is gettered by the sputtered Ti species deposited at the walls of the reactor (and on the film). Moreover, the ion-induced sputtering mechanism prevents the target from being oxidized. In this region, the deposition rate is almost unaffected by the oxygen and the film stoichiometry smoothly increases with the amount of oxygen in the reactor.
- Transition region (around 2 sccm of oxygen in Fig. 3.5): Here, the amount of oxygen in the plasma is high enough to start to oxidize the first monolayers of the target in a phenomenon widely known as target poisoning. The ion bombardment causes the sputtering of not only Ti but also O atoms, decreasing the sputtering rate of Ti atoms and the gettering of oxygen by the walls. Consequently, the deposition rate rapidly falls while the partial pressure of the oxygen rapidly increases with the oxygen flux. It is in this region where the film stoichiometry rapidly increases and usually reaches the full saturation with oxygen.
- Poisoned mode (above 2.5 sccm in Fig. 3.5): The amount of oxygen in the plasma is high enough to fully oxidize the first monolayers of the target material, and the target is said to be fully poisoned. This causes a drop in the sputtering and deposition rates of the titanium and the minimization of the gettering process. In this region, the film stoichiometry remains fully saturated with oxygen but the deposition rate falls considerably in comparison with that in the metallic mode.

For decreasing oxygen fluxes, a similar reverse behavior can be obtained although the transition region is now found for a lower oxygen flux of 1.5 sccm, favoring the appearance of a hysteresis curve in Fig. 3.5. The shift of this transition region when the oxygen flux is decreased is mediated by the following factors: (i) the existence of fully oxidized layers in the target in contact with the plasma makes the propagation of electromagnetic power less efficient and, thus, the ion bombardment less energetic, and (ii) a fully oxidized target does not favor the sputtering of Ti atoms and the gettering of oxygen at the walls. Consequently, the only way to decrease the amount of oxygen in the plasma and deoxidate the target is to decrease the oxygen flux even more.

As shown in Fig. 3.5, the deposition rate of Ti atoms falls by a factor of ~15 when shifting from the metallic to the poisoned mode of discharge and, as mentioned before, represents a clear drawback of the r-MS technique. In fact, much research has been carried out in the last decades to avoid target poisoning and keep a high deposition rate when growing fully stoichiometric compound thin films. For further information on this issue, please see Ref. [10].

3.2 Plasma-Assisted Deposition of Porous Nanocolumnar Thin Films

3.2.1 The Oblique Angle Geometry

MS deposition at oblique angles (MS-OAD) is a variation of the classical version of the technique, in which the deposition species arrive at the substrate along a preferential oblique direction. From a fundamental point of view, this geometry induces the appearance of the so-called surface shadowing processes, responsible for introducing particular growth dynamics and the development of nanocolumnar structures. In contrast to other fundamental processes during growth that tend to smooth out the surface of the film or minimize the surface energy, the shadowing process is a nonlocal mechanism that tends to destabilize the surface by promoting the preferential growth of taller surface features over shorter ones [5]. Its influence is briefly described in Fig. 3.6. In the first stages of growth, the deposition species arrive at the substrate along a preferential oblique direction, being deposited at random locations and forming small islands (Fig. 3.6a). However, as soon as some height fluctuations appear on the surface, some regions will be "shadowed" by these tall features,

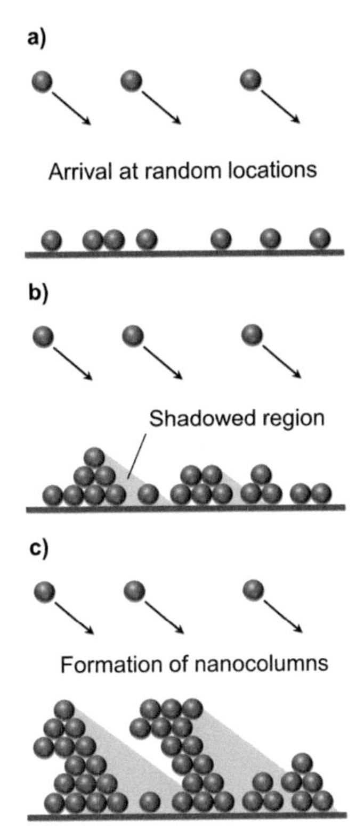

Figure 3.6 Illustration of how nanocolumns are formed when the deposition species arrives at the substrate at oblique angles. The deposition takes place at random locations (a) until atoms start to pile up and form mounds (b). These mounds, due to the oblique angle incidence of the deposition species, cast a "shadowed region," where no species can be directly deposited. This makes the taller surface features grow faster than the lower ones, causing a competitive phenomenon among them and the formation of tilted nanocolumnar structures (c).

that is, no deposition species may directly land at that particular location, where the growth is inhibited (Fig. 3.6b). This introduces a competitive process by which the taller a mound is, the longer is the shadowed region behind it, and the faster it develops, ending up in the formation of porous nanocolumnar structures (Fig. 3.6c). The tilt angle of the columns with respect to the normal of the surface is usually called β, and its relation with the angle of incidence of the deposition species has been a relevant research topic in the last decades. As a general rule, the more oblique the angle of incidence is, the higher is the value of β, the columns being typically separated a distance defined by the length of the shadowed regions. Yet, there are numerous factors defining not only the intercolumnar distance but also the inner porosity of each column, such as the energy of the species or the broadening of the angular distribution of the deposition species.

According to the preceding discussion, the oblique angle geometry can be implemented in MS deposition by (i) promoting a ballistic deposition regime, where the deposition species arrive at the film surface along well-defined directions, and (ii) promoting the arrival of the deposition species along oblique directions with respect to the substrate. The first condition can be achieved by operating at low thermalization degrees, that is, at low pressures and/or low target-film distances, while for the second condition, there are numerous configurations, the most popular being achieved by mounting the film on a rotatable substrate holder so that the target and film surfaces are not parallel (see Fig. 3.7). However, there are other geometries and operational conditions allowing MS deposition at oblique angles by means of obstacles or particle collimators, for instance.

Tilting the substrate holder with respect to the target a certain angle causes the ballistic species to arrive at the film surface along certain preferential directions. However, the fact that the species are not sputtered from a punctual source, but from the racetrack, introduces some relevant effects. The first one relates to a certain difference between the angle of incidence of deposition species and the tilt angle of the substrate [11]. As illustrated in Fig. 3.7, for a certain tilt angle of the substrate, α, the polar angle of incidence, θ, will be $\alpha + \Delta$ and $\alpha - \Delta$ for the species sputtered from the top part and the bottom part of the racetrack, respectively, with $\Delta = \tan^{-1}(R/L)$, where R is the radius of the racetrack, while those sputtered from the sides of the racetrack would arrive with a polar angle Δ. Typical values of Δ are between 15° and 20°. In this way, and as illustrated in Fig. 3.8, even when all deposition species are ballistic, the distribution of the polar angles of incidence when the substrate is tilted would have a wide distribution from $\theta = \Delta$ to $\theta = \alpha + \Delta$. In Fig. 3.9 we see the result of a simulation of the angular incidence of Ti species at low pressures when $L = 7$ cm and $R = 2.5$ cm, that is, $\Delta \approx 20°$ for different values of α. There, we can see a ring-shaped distribution, resembling the racetrack, that shifts to higher polar angles as we tilt the substrate. Remarkably, for $\alpha = 70°$, a second relevant effect of tilting the substrate can be noticed: the racetrack might be partially covered by the substrate when the tilt angle surpasses the threshold $90° - \Delta$, inhibiting the ballistic contribution of the bottom part of the racetrack. This means that the substrate can be tilted to an angle up to $90° + \Delta$ (in Fig. 3.9 we also include the angular distribution when $\alpha = 100°$) and still make a ballistic contribution to the film growth from the upper part of the racetrack. This situation is apparently beneficial as the source of sputtered species is now rather punctual and well collimated. However, we must take into account that the contribution of thermalized sputtered atoms remains the same, no matter the tilt angle of the substrate, and when we cover some parts of the racetrack, we decrease the amount of ballistic species, which would be detrimental in terms of angular definition and deposition rate.

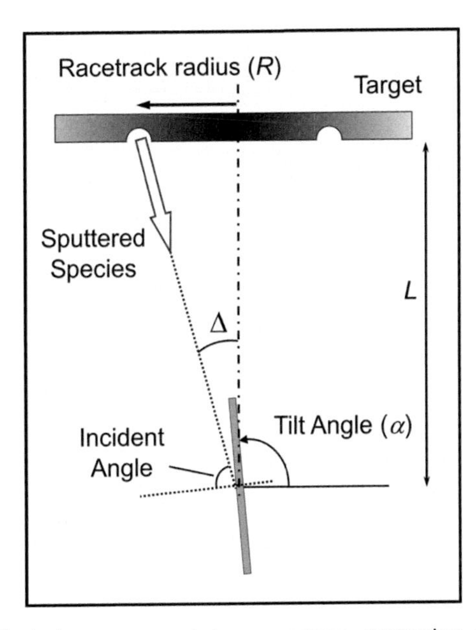

Figure 3.7 Typical arrangement in magnetron sputtering depositions at oblique angles. The incident angle of the deposition species depends on the tilt angle of the substrate, α, and the configuration of the racetrack by means of the angle Δ.

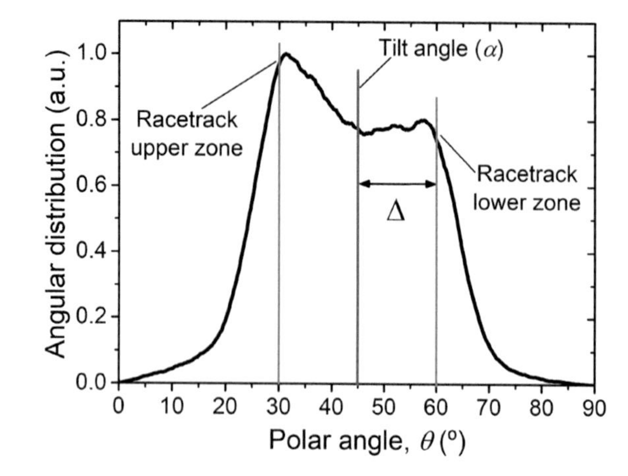

Figure 3.8 Typical polar angle distribution of the deposition species. While the tilt angle of the substrate is $\alpha = 45°$, the angular distribution significantly widens due to the circular shape of the racetrack, with $\Delta \approx 15°$. Note that tilting the substrate makes the film closer to the upper part of the racetrack, and that is why the peak at $\vartheta \approx 30°$ is higher than that at $\vartheta \approx 60°$.

3.2.2 Process-Control and Growth Mechanism

To grow nanocolumnar porous thin films, the surface shadowing processes must dominate the nanostructuration of the film over other processes that tend to smooth out the surface and/or reduce the film porosity. For this, the oblique angle configuration is implemented under conditions where thermally induced mobility is minimized, in other words under conditions where the Thornton diagram indicates that these processes play a minor role, that is, when $T_f/T_m < 0.25$, where T_f is the film temperature during growth and T_m the melting temperature of the material [4, 12]. When working in this zone of the diagram, and in the absence of plasma-film interaction during growth, the key experimentally controllable quantities defining the nanocolumnar structure are the tilt angle of the substrate with respect to the target, the target–film distance, and the pressure of the gases in the plasma. Next, we summarize the main influence of these variables on the film nanostructure:

- Pressure of the gases in the plasma, p_g: As mentioned above, gas pressure directly affects the thermalization degree of the deposition species when they arrive at the deposition surface.

The dependence is such that the lower the working pressure, the more ballistic the deposition species would be, its lower limit being defined by the minimum pressure necessary to maintain the sputter plasma active. In this way, the angle of incidence would be better defined at lower pressures. Moreover, the kinetic energy of the deposition species is also defined by this pressure. As we will comment in Section 3.2.4, hyperthermal processes associated with the deposition species with a kinetic energy above the bond energy of surface species may produce certain atomic relaxation processes and alter the film nanostructure. Furthermore, higher pressures would involve an increasing contribution of thermalized species in the deposition, characterized by an isotropic angular distribution in the gas phase and typical energies in the thermal range.

- Target-film distance, L: Like p_g, this quantity affects the growth of the film through the value of the thermalization degree, although there are some particular aspects that have to be taken into account. The minimum value of L is defined by the necessity to maintain the film outside the plasma zone. Indeed, the most intense part of the plasma is mainly confined to within a few centimeters from the target, depending on the electromagnetic power employed to maintain it or whether the magnetic confinement of the electrons is balanced or unbalanced. In fact, if the film is placed inside the plasma, there can be some relevant processes, such as thermal heating or ion/electron impingement on the film, that tend to smooth out the film surface, thus reducing the porosity. Moreover, the value of L also determines the value of the angle Δ, of special relevance in terms of the angular distribution of the ballistic deposition species.
- Tilt angle of the substrate, α: As mentioned above, the tilt angle of the substrate mediates the value of the incident angle distribution function of the ballistic species, and its value may vary from 0° to 90° + Δ. From a practical point of view, angles of incidence below ~60° produce rather compact film nanostructures, while the columnar breakdown (the sudden appearance of columns) takes place for angles between 60° and 70°.

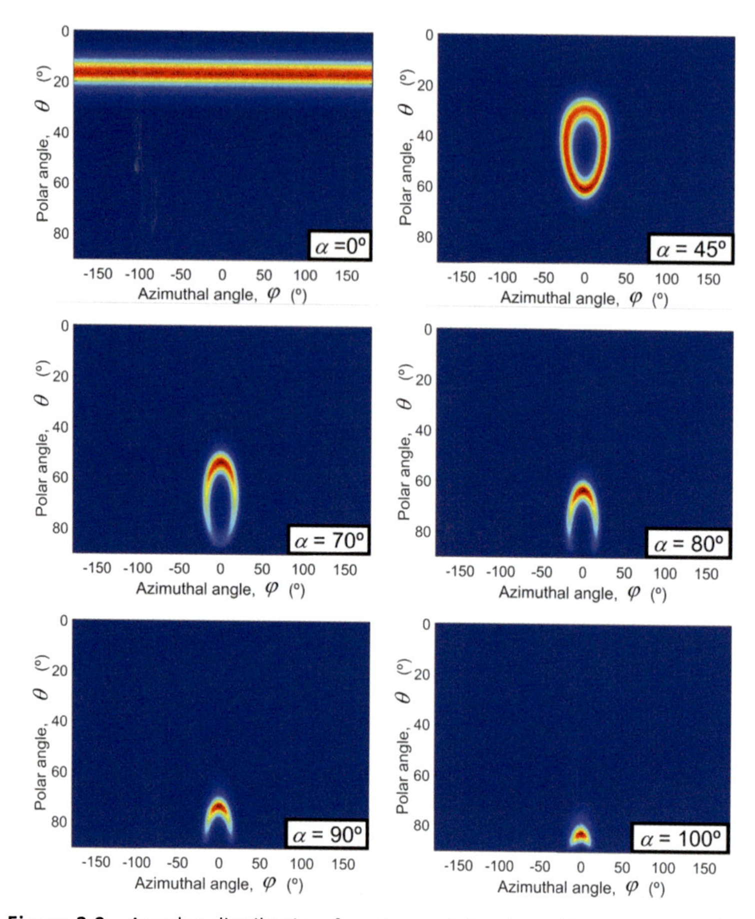

Figure 3.9 Angular distribution functions of the deposition species at low pressures and for different values of the tilt angle of the substrate. When $\alpha = 0°$ (classical configuration) the polar angle of incidence (θ) is well defined, around 15°–20°, independent of φ. On tilting the substrate ($\alpha = 45°$), and due to the fact that the angles are associated to the reference system (substrate), the circular shape of the racetrack can be made out. For $\alpha = 70°$ the lower part of the race track stops contributing to the film growth. This effect is enhanced for $\alpha = 80°$, $\alpha = 90°$, up till $\alpha = 100°$, where only the top part of the racetrack contributes to the growth of the film, acting as a nearly punctual source.

From an experimental point of view, the three quantities mentioned above are usually tuned to achieve different columnar structures whose particular features also depend on the particular geometry of the reactor, material deposited, etc. A theoretical and experimental analysis of the influence of Ξ and α on the film nanostructure was carried out for gold thin films, identifying four generic nanostructures depending on the operating conditions as illustrated in Fig. 3.10:

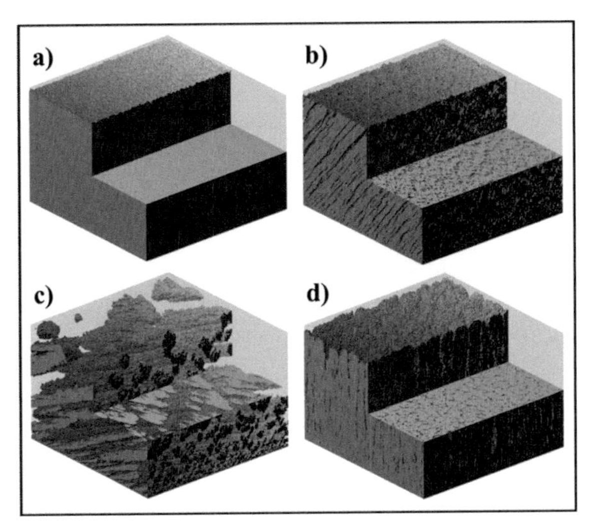

Figure 3.10 Typical nanostructures found at oblique angles. At low incident angles and low thermalization degrees, compact nanostructures are typically found (a). When the tilt angle of the substrate increases, the nanostructure develops tilted mesopores that percolate from the top to the bottom of the film (b). At even higher angles, typical tilted columnar structures emerge, giving rise to a highly porous film. Finally, at high thermalization degrees, irrespective of the tilt angle of the substrate, vertically aligned and coalescent sponge-like structures define the film nanostructure (d). Reprinted from Ref. [13]. © IOP Publishing. Reproduced with permission. All rights reserved.

- When $\Xi << 1$ and $\alpha \lesssim 60°$, a compact nanostructure is found with little or no pores (Fig. 3.10a).
- When $\Xi << 1$ and $60° \lesssim \alpha \lesssim 80°$, the film nanostructure is still rather compact but now numerous voids are apparent percolating from the very top of the film to the bottom that introduce some open porosity into the films. Overall, the film resembles a highly packed and tight nanocolumnar thin film where columns are highly coalescent (Fig. 3.10b).
- When $\Xi << 1$ and $\alpha \gtrsim 80°$, the film possesses some similarities with respect to the one before, but now the voids are larger

and the columns clearly develop independently. In this region, the more tilted the substrate, the higher the value of β (Fig. 3.10c).

- When $\Xi >> 1$, no trace of nanocolumns is found no matter the value of α, which is coherent with the thermalized nature of the deposition. Here the film nanostructure is formed by vertically aligned sponge-like structures whose pores are well connected and percolate from the top to the bottom of the film (Fig. 3.10d).

Remarkably, when $\alpha \gtrsim 80°$ and Ξ increases, columns become more vertical and closer to each other and start to develop certain internal porosity until $\Xi >> 1$, where each column turns into a vertically aligned porous and highly coalescent sponge-like structure.

3.2.3 Effect of the Kinetic Energy of the Deposition Species on the Nanocolumnar Growth of Thin Films

In the previous section, the influence of geometrical constraints and the progressive thermalization of the deposition species have been analyzed. There are, however, two important issues that will be developed in this and the following sections: the influence of the incident kinetic energy of the deposition species and the likely interaction between the film and the plasma during growth.

In terms of kinetic energy, one can conclude that the ballistic species possess a much higher kinetic energy when being deposited than thermalized or partially thermalized species. As mentioned in Section 3.2.2, the energy distribution peaked at $U/2$ but had a long tail for higher energies. This means that there can be a significant number of ballistic species with energies above U. Assuming the film has the same composition as the target, this means that there is an important number of ballistic species that can break the bonds of the surface atoms and cause some mobility processes. This entails the existence of an important relaxation process in the material's network, not related to thermal activation processes, that is, of a hyperthermal nature. This kinetic energy–assisted relaxation mechanism introduces new relaxation channels directly related to the ballistic deposition of the species (see Fig. 3.11a).

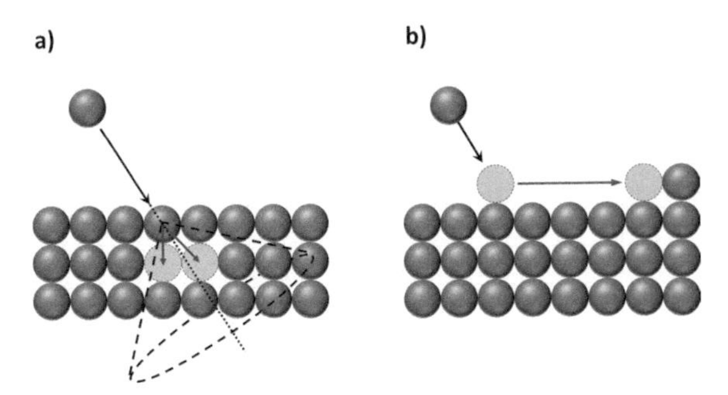

Figure 3.11 Typical hyperthermal relaxation processes induced by the kinetic energy carried by the deposition species in the gas phase. Kinetic energy–induced relaxation by binary collision (a) and biased diffusion (b).

A second process accounts for the possibility that when high-energy deposition species arrive at the film surface, they bond to it, neutralizing the perpendicular component of momentum but keeping the parallel one (see Fig. 3.11b). As a consequence, the deposited species may slide over the surface until they come to a stop at a position far from the landing point. This process, called "biased diffusion," is responsible for a net flow of the deposition species over the surface of the film, which together with the kinetic energy–assisted relaxation mechanism may affect the growth of the film. Of course, there are more hyperthermal processes that could be responsible for altering the film's nanostructure, although in this chapter, we focus on these two as they have been studied in detail in relation with MS depositions.

In Ref. [13] a study on the growth of Ti thin films at oblique angles was made under ballistic deposition conditions, demonstrating the key role of the two above-mentioned hyperthermal processes. In fact, it was demonstrated by means of fundamental experiments and simulations that these processes made the columnar nanostructure rather vertical and independent of α. In Fig. 3.12a we show a simulation of a Ti thin-film grown under conditions where $\Xi << 1$ and $\alpha = 85°$, including these two hyperthermal processes, next to a cross-sectional SEM image of the film (see Fig. 3.12b), where we can see the existence of rather vertical columns, as well as the good agreement between the results of the model and the experiments. In Fig. 3.12c, we show the result of the simulation when the influence of these two processes was turned off and where a more classical structure formed by thinner and more tilted columnar structures is apparent. This simulation would be realistic, therefore, when the kinetic energy of the deposition species is low enough to inhibit hyperthermal deposition species. From an experimental point of view, this can be achieved by using a rather different growth technique, the so-called electron beam–assisted evaporation at glancing angles, where a solid material is sublimated in a vacuum reactor by the impingement of an electron beam and where the geometry makes that deposition species arrive at a surface along an oblique direction. In this case, the typical kinetic energy of the deposition species is ~0.1 eV, and hence no hyperthermal processes should be expected. The cross-sectional image of the film for an incidence of 80° is presented in Fig. 3.12d and reveals the good agreement between the model without hyperthermal processes and the experiment, demonstrating that kinetic energy–induced relaxation mechanisms and biased diffusion processes must be considered in MS depositions at oblique angles.

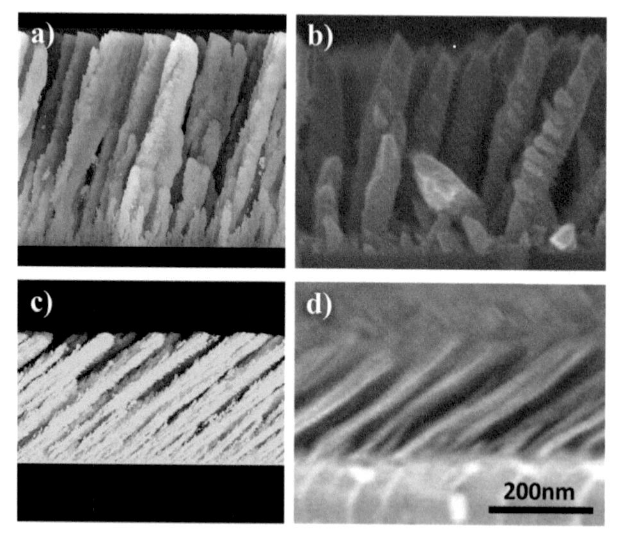

Figure 3.12 Simulation of the nanostructure of a Ti thin film when hyperthermal processes are considered (a). SEM image of a Ti thin film grown at oblique angles at low pressures (b). Simulation of the same nanostructure shown in (a), but inhibiting the influence of hyperthermal processes. SEM image of Ti thin film deposited by evaporation, with the same angle of incidence of the deposition species as in (b). Reprinted from Ref. [14]. © IOP Publishing. Reproduced with permission. All rights reserved.

3.2.4 Influence of Plasma–Nanocolumnar Film Interaction during Growth

In MS depositions, the film is placed a few centimeters away from the target, where the plasma is responsible for the sputtering mechanisms. Yet, and even though the plasma is confined next to the target, some energetic species can reach the film and affect the nanostructural development of the layers. This is a current topic of research, and there are still many unknown effects related to the impingement of energetic species on the film. These are listed next.

- Positive ions: The plasma contains numerous positively charged atomic or molecular species that may impinge on the film during growth, induce mobility of the deposited species, and alter the nanostructure and film porosity. In this regard, the movement of positive ions in the plasma is defined by the density gradients (diffusive term) and the direction of the electric field (mobility) in the plasma. If the substrate is not electrically biased, the ions reaching the film will have a kinetic energy equal to the potential drop from the plasma potential to the potential of the substrate (0 if it is grounded or the floating potential). Here what is important is not only the energy flux from the ions to the film but also the net current of ions and the ion energy. In normal conditions, if positive ions are required to play a role in the development of the film nanostructure, a substrate electrical bias would have to be applied whose characteristics would depend on the electrical nature of the film and the substrate [14].
- Negative ions: Although the electric field a few centimeters away from the target is pointing toward the substrate and, hence, negative ions are dragged toward the bulk of the plasma, there is an important mechanism responsible for producing high-energy negative ions that could affect the porosity and nanostructure of the film. This occurs, for example, when using oxygen to change the film stoichiometry, a situation that we will use as an example to illustrate this phenomenon. In this case, atomic oxygen may get very close to the target and, just before becoming chemisorbed, it may recombine with a secondary electron, forming a negative ion within the plasma

sheath. As a consequence, it will be accelerated away from the target, gaining a kinetic energy equal to the target's potential (on the order of a few hundred electron volts), pass through the plasma, and eventually impinge onto the film, where such an impact might cause similar phenomena as ions on the target surface [15].

- High-energy reflected neutrals: Similar to negative ions, a high energy positive ion in the plasma sheath accelerated towards the target may experience a charge transfer mechanism with a low energy neutral species, resulting in an ion with low energy and a neutral species with high energy that may go back to the plasma bulk. In this way, the impingement of a high-energy neutral on the film surface can induce processes similar to those induced by the positive ion on the target, altering the film's nanostructure.
- Radiation from the plasma/target: In addition to ions and electrons, the plasma is formed by numerous species in excited states that emit radiation upon electronic decay to their fundamental levels. The emitted light might be absorbed at the film surface and cause local excitations and induce certain mobility [16].
- Excited species in metastable states: In the same way that the plasma creates numerous species in excited states that may radiate, it might also create excited species in metastable states that might diffuse toward the film and leave its energy on the surface, causing similar processes.
- Others: In this category, we consider numerous species that might be produced in the plasma and that could affect the film growth. For instance, depending upon the composition of the plasma, numerous reactive species could be generated that might get to the film surface and chemically etch some species from it, releasing some energy and altering both the film chemical composition and its nanostructure [16].

The list of energetic species that might influence the film nanostructure is large, as the plasma may operate as a source of many of them. The general outcome of the interaction between film and energetic species is the enhanced mobility of the deposited atoms, an effect that goes against the expected results of surface shadowing mechanisms at oblique incidence, promoting the chemisorption of the deposition species within shadowed regions and diminishing the efficiency of the formation of nanocolumnar structures. Therefore, and even though in most cases the flux of energetic species toward the film is low enough to not cause appreciable modifications in the film nanostructure, it is usually recommended that the substrate be placed far from the plasma to ensure a high porosity of the films grown at OAD geometries.

3.2.5 Reactive Magnetron Sputtering at Oblique Angles

In Section 3.1.4 we described most important principles in r-MS depositions and showed that a certain amount of reactive gas may be introduced in the plasma discharge to alter the film chemical composition. This changes the nature of the sputter plasma and hence the sputtering mechanism and the transport of species toward the film. Anyway, the main principles governing film growth and nanocolumnar development are the same as mentioned above, even though the presence of a reactive gas may prompt the presence of new energetic or reactive species that could alter the film composition and nanostructure. Yet, the amount of reactive gas employed in normal conditions is much lower than that of the nonreactive gas, and usually no relevant changes are apparent in the film, except for the variation in its chemical composition.

There are several aspects of r-MS deposition that can be improved when operating at oblique angles. In fact, in Section 3.1.4, we saw that most undesired phenomena in this type of depositions relate to target poisoning and low deposition rates. These effects can be minimized by using a methodology based on an OAD geometry. To develop its principles, we will follow the same example as in Section 3.1.4 on the r-MS deposition of TiO_x thin films by an Ar/O_2 plasma, which is presented in Fig. 3.13. In this figure, we saw the relevant change in the deposition rate in a classical configuration between the metallic and poisoned modes. The only difference between both conditions was an increase in the oxygen flux in the reactor from 2.5 sccm to 2.8 sccm. However, the metallic mode promoted the deposition of a substoichiometric film up to $x \approx 1.9$ at a relatively high growth rate of $r \approx 25$ atoms/nm^{2}s, while the poisoned mode allowed the growth of a fully stoichiometric TiO_2 thin film with a growth rate of $r \approx 1.6$ atoms/nm^{2}s, that is, a drop of ~95% in the growth rate of

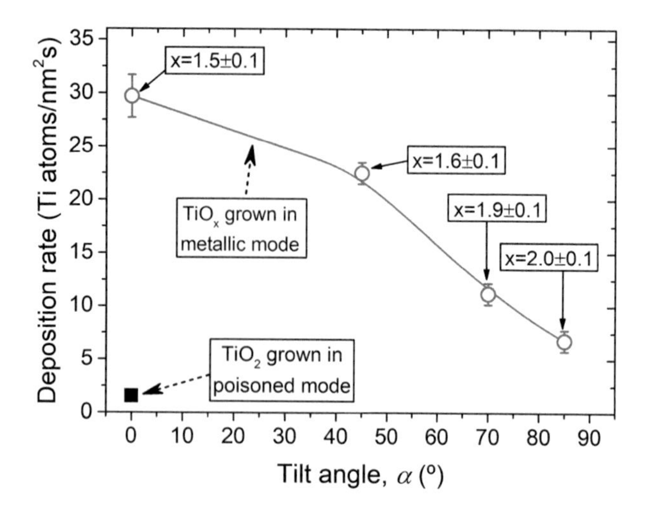

Figure 3.13 Change in the stoichiometry of films with a composition TiO_x for given deposition conditions and for different tilt angles of the substrate with respect to the target.

the films. To minimize this decrease and to grow fully stoichiometric thin films at higher growth rates, a simple strategy based on OAD geometry can be applied. The main idea behind this procedure relies on the relatively high deposition rate obtained in the metallic mode and the fact that a decrease in the deposition rate of Ti atoms under those conditions is of 5%–10%, while keeping the impingement rate of oxygen species constant would lead to fully stoichiometric films with growth rates only slightly below, avoiding the drastic fall when operating in the poisoned mode. This can be achieved by tilting the substrate under metallic mode conditions, which would result in a reduction in the deposition rate of sputtered ballistic species due to a change in the area factor, while the impingement rate of oxygen species on the substrate shall be independent of the tilt angle of the substrate due to their isotropic angular distribution in the plasma gas. In Fig. 3.13, we show the change in the film deposition rate and stoichiometry when the substrate is tilted with respect to the target and how a stoichiometric film can be deposited at higher deposition rates than in the poisoned mode (see also Fig. 3.5). However, the main difference between the film grown at OAD and that in the classical geometry is its porosity. While the former possesses a nanocolumnar porous nanostructure, the latter is rather compact. However, as seen in previous sections, film porosity can be lowered or removed by inducing the impingement of plasma ions or other energetic species on the film during growth, for instance, by electrically biasing the film. In this way, OAD represents a good approach not only to grow porous thin films but also to minimize the main drawbacks of operating in reactive conditions, avoiding target poisoning and the growth of fully stoichiometric compound layers at relatively high deposition rates.

3.2.6 Growth of Nanocolumnar Thin Films on Patterned and Rough Substrates

In addition to the above-mentioned process-control variables, the film morphology can be strongly modified by the presence of patterns on the substrate. Indeed, and as we have seen in the previous sections, the main mechanism responsible for the appearance and development of the nanocolumnar structures is surface shadowing, which at oblique incidences promotes the growth of taller surface features in detriment of lower ones. In this way, when the substrate is flat, height fluctuations associated with the random arrival of deposition atoms at the substrate represent the main mechanisms responsible for the formation of mounds and the subsequent formation and development of nanocolumns. However, when the substrate shows a specific intrinsic roughness or possesses a specific height pattern, the tallest parts would define the regions where mounds would likely nucleate and promote the formation of nanostructures on top. In this case, depending on the film thickness, three different growth stages can be differentiated. These are shown in Fig. 3.14, where a nonperiodic ion-sculptured rippled substrate has been employed to grow a SiO_2 thin film at an oblique angle [17]:

- At low film thicknesses, the film preferentially grows on the tallest parts of the substrate, defining a substrate-driven growth regime where the film topography is defined by the substrate pattern. In this region, and according to Fig. 3.14, the surface correlation length of the film is defined by that of the pattern.

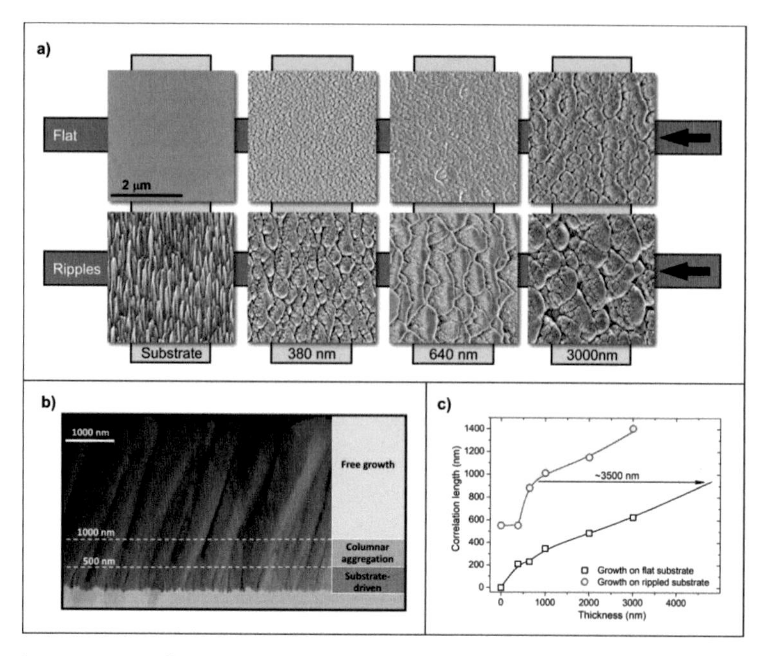

Figure 3.14 Influence of substrate patterns on the nanocolumnar growth of thin films by magnetron sputtering at oblique angles. (a) Top view SEM image of the films deposited on a flat Si substrate and on an ion-induced patterned substrates, for increasing thicknesses. It is apparent how a columnar structure seem to follow the substrate pattern for low film thicknesses and how it is progressively lost for increasing thickness values. (b) Cross-sectional TEM image of the film grown on the patterned substrate, where three different stages of growth can be differentiated. (c) Value of the correlation length defining the film surface in the direction perpendicular to the patterns, where the three stages of growth are clearly depicted and where it is shown that the large columnar structures formed in the free growth regime (high thicknesses) grow similar to the nanocolumns on flat substrates. Reprinted from Ref. [18] with permission from John Wiley and Sons, Copyright (2018).

- The substrate-driven growth regime dominates the film development until it reaches a certain critical thickness, the so-called oblivion thickness, above which larger columnar structures start to take over smaller ones, forming large structures. Here the surface correlation length of the film deviates from that of the substrate pattern in a so-called columnar aggregation growth regime.

- Finally, for large film thicknesses, the film topography tends to resemble that of an equivalent thicker film deposited on a flat substrate in a so-called free-growth regime.

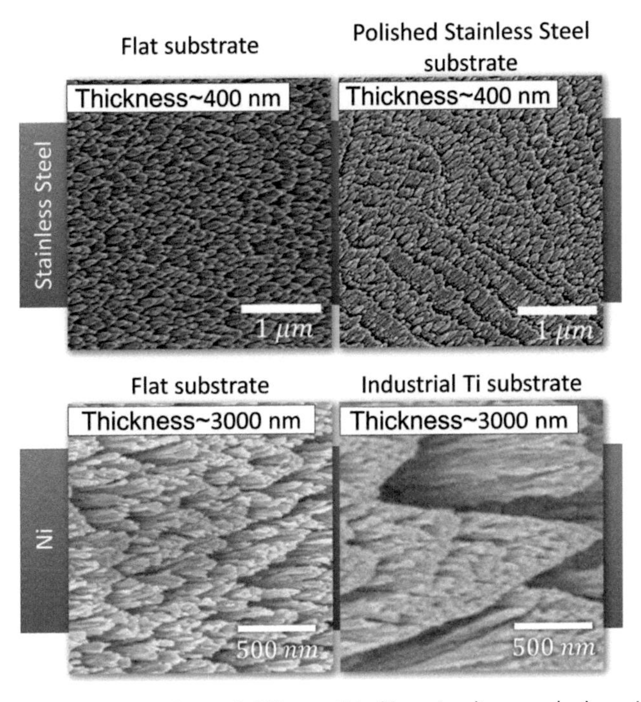

Figure 3.15 Top SEM views of different thin films simultaneously deposited at oblique angles on flat and rough substrates. On the top, the nanostructure of a stainless-steel film deposited on a flat Si substrate and on a polished industrial stainless-steel substrate. For a film thickness of 400 nm we see the existence of columnar fronts in the latter case, associated with the polishing process and the existence of patterns on the film surface. Below, Ni thin films were deposited at oblique angles on Si and on industrial Ti with a thickness of 3000 nm. Here, we notice that in the latter case, the large structure in the image resembles the typical shape of a magnified column.

The specific film thickness above which we find a transition from one regime to the following one would depend on the deposition conditions as well as on the specific features of the substrate pattern. However, this explains quite accurately the main effect of using a rough substrate instead of a flat one when depositing a film at oblique angles. In Fig. 3.15 (top), we show images of two stainless steel thin films grown simultaneously, one on a flat substrate and one on a polished substrate, suggesting that the nanocolumns follow the pattern created by the polishing procedure on the substrate in a typical substrate-driven growth regime. In Fig. 3.15 (bottom), we can see two Ni films simultaneously grown on a flat industrial Ti substrate and an unpolished one. There we can see that the topography of the film grown on the rough substrate resembles that of the film grown on the flat one, but with much larger structures, thus reproducing the topography of a much thicker film deposited on a flat substrate in a typical free-growth regime. Even though these results could be extrapolated to the growth of thin films with specific patterns on different scales, most experimental results on this topic have been obtained following the classical arrangement of the MS technique and not the oblique angle configuration. The generalization of the scheme presented above is a new and exciting topic of research that will be developed in the near future.

3.3 Nanostructure-Related Applications of Porous Nanocolumnar Thin Films

The nanocolumnar and porous microstructures of thin films prepared by MS offer a series of possibilities that can be of much interest in improving their performance for specific applications. Several factors may contribute to this improvement of performance, paving the way for the use these thin films in a large variety of applications. We will look at applications derived from the following properties of MS thin films prepared under specific conditions: (i) the effect of high (homogeneous) porosity for thin films prepared at relatively high plasma pressures, (ii) the nanocolumnar microstructure characteristic of MS-OAD thin films formed by nanocolumns, and (iii) the nanostructuration phenomena appearing in thin films deposited onto patterned substrates. In this section, we will discuss these features in relation with some applications that, directly stemming from them, demonstrate the use of the MS technique in the preparation of porous and nanostructured thin films. In the next three sections, we will comment in more detail on each one of these nanostructuration processes, presenting specific examples of applications in various technological areas.

3.3.1 Porous Magnetron Sputtered Thin Films

Thornton [4, 12] had already pointed out that increasing the plasma gas pressure during MS deposition modifies the nanostructure of the thin films in such a way that the column width decreases and the interface area between them maximizes. According to what has been discussed in Section 3.1.3 high plasma gas pressures during MS deposition produce effective thermalization of the sputtered particles, which then tend to arrive at the substrate in randomized directions. A high thermalization regime of deposition particles can be achieved by increasing the working pressure and/or by increasing the distance between target and substrate. It is remarkable that at a high thermalization degree of sputtered particles, the microstructure of MS thin films may resemble that of plasma-enhanced chemical vapor deposited films, characterized by vertical and sponge-like nanocolumns [18]. In previous publications, we have studied this microstructure and linked its development with what we denoted as an isotropically directed deposition flux. The isotropical character of this deposition was effectively confirmed by the formation of a thin film not only on the side of the substrate facing the magnetron target, but also on its back (cf Fig. 3.16).

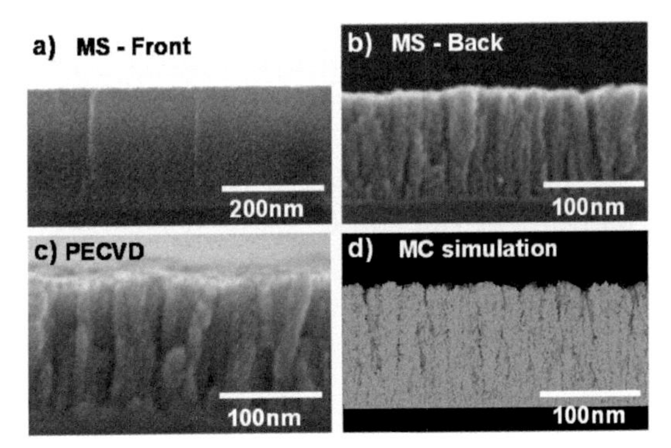

Figure 3.16 Cross-section micrograph of TiO_2 thin films prepared by MS (a, b) and PECVD (c). MS deposition (a) usually gives rise to more compact thin films than PECVD (c). However, under high thermalization degree conditions, as on the back of the substrate not exposed to the target (b), the film adopts a more open microstructure, very similar to that of PECVD films. MC simulation (d) under the assumption of considerable thermalization reproduces well the microstructure of the PECVD and MS-back film. Reprinted from Ref. [19], with the permission of AIP Publishing.

An additional effect of a high thermalization degree of deposition particles during MS is the increase in the porosity of the films in the form of small micropores (i.e., micropores with sizes smaller than 2 nm, according to the IUPAC [19]). These porous thin films have found applications in different domains where a high surface area in contact with the medium is a key feature. Herein, we will briefly discuss potential applications of this porous microstructure in the following domains: electrochromism, conductometric gas sensor films, and porous membranes.

- Electrochromic and related thin films: An electrochromic device deposited onto a flat glass substrate consists of a cathode and an anode in the form of thin films separated by a, generally solid, electrolyte. A typical cathode material is WO_3, which changes its optical absorption properties when the W^{6+} cations present in the oxide are partially reduced, and cations from the electrolyte are incorporated into its structure according to the following reversible electrochemical reaction [20]:

$$WO_3 + xe^- + xM^+ \rightarrow WO_3\,(xM), \tag{3.1}$$

where M^+ is a small size cation, generally Li^+. The coloration efficiency of this cathode depends on its capacity to incorporate cations from the electrolyte, while the kinetics of the change from the bleached state (WO_3) to the colored state (i.e., $WO_3(xM)$, where a number of tungsten ions appear in a W^{n+} oxidation state with $n < 6$), strongly depends on the porosity of the films. These two characteristics are directly correlated with the porosity of the films, which, in general, makes accessible the inner part of the thin-film electrode to the M^+ cations of the electrolyte and favors their diffusion into the interior of the film. MS is one of the most common methods for the manufacture of WO_3 porous thin films, particularly at industrial scale. A common strategy to enhance the porosity in these MS thin films is the use of high-thermalization-degree conditions for the deposition. Although the majority of the works in the scientific literature address the problem of controlling the composition of the film prepared under a reactive MS regime and how adjusting the oxygen particle pressure can be used to get stoichiometric thin films, there are also a few studies that refer to the dependence of film porosity on the plasma gas pressure and its effect on coloration dynamics. For example, in Ref. [21], the authors studied in a systematic way the effect of working pressure on the optical and electrochromic properties of WO_3 thin films prepared by MS. They found that the amorphous thin films prepared at high pressure exhibited a lower density, faster coloration dynamics, and good reversibility, thus supporting that processes according to Eq. 3.1 were promoted thanks to the high porosity of the films prepared under these conditions. In a more systematic way, Salinga et al. [22] showed that the decrease in the deposition rate (due to scattering of sputtered atoms with the plasma gas) and film density observed at pressures higher than 2 Pa could be accounted for by the Keller–Simmons model (see Section 3.1.3). As a result, they found a strong dependence of coloration behavior and switching rate of the WO_3 thin films on the thermalization processes of sputtered particles. Other approaches to prepare porous and nanostructured thin films for an enhanced electrochromic response rely on the use of an MS-OAD configuration, a methodology well known to render porous and nanostructured thin films and that has been utilized in several papers dealing with the electrochromic behavior of WO_3 and other oxides [20–23].

A similar concept involving an increase in the pressure during MS deposition to enhance the thermalization degree during deposition was followed for the preparation of color sensors of hydrogen consisting of Pt-loaded WO_3 thin films. These were deposited on oxidized Si substrates under various discharge gas pressures at room temperature using reactive MS. The films became more porous as the plasma gas pressure increased leading to a decrease in the film density and an increase in the effective surface area [24]. In parallel, these films depicted an enhanced coloration efficiency when they were exposed to traces of hydrogen (note that the coloration mechanism in this case is different from that in Eq. 3.1 and entails a chemical reaction of the type $WO_3 + (3 - x)H_2 \rightarrow WO_x + (3 - x)H_2O$).

- Gas sensor films: These films usually consist of semiconductor oxides such as ZnO, SnO_2, TiO_2, and WO_3 that change their electrical conductivity if the concentration of the free carriers in the conduction band increases/decreases when they are brought in contact with a reducing/oxidizing compound at relatively high temperatures. To enhance the changes in electron conductivity, even in the presence of very low concentrations of the compound to be detected, it is common to incorporate metal nanoparticles to catalyze its decomposition, enhancing the interaction degree with the semiconductor oxide. The sensing mechanism relies on a process that, for the particular case of reducing hydrocarbons, can be written as

$$MO_x + C_xH_y \rightarrow MO_{x-z} + C_xH_y. \quad (3.2)$$

This process entails the reduction of a minority concentration of M^{x+} cations and the appearance of defect states close to the Fermi level of the semiconductor oxide. This, upon thermal excitation, produces an increase in the free electron concentration in the conduction band of the semiconductor oxide and an increase in the electrical conductivity of the system.

A condition for an enhanced and fast electrical response of the semiconductor oxide is to maximize the area of contact of the thin-film sensor with the medium (i.e., the extent by which the process in Eq. 3.2 is displaced to the right). For example, ZnO is a good sensor material for the detection of ethanol, ammonia and other compounds and, in the form of a porous thin film, can be efficiently prepared by reactive MS. To increase the porosity and exposed surface area of the ZnO thin films, a common strategy is to change the plasma gas pressure during deposition. This was the effect reported by Fairose et al. [25], who found a higher sensitivity for ammonia detection at room temperature for the thin films prepared at relatively higher pressures. Although increasing the value of this processing parameter usually produces an increase in sensitivity, in some cases undesirable surface and size effects may be associated with an increase in the plasma gas pressure during deposition. For example, in the work of Kwoka et al. [26], the ZnO thin films prepared at a maximum pressure presented a nanostructure formed by agglomerated nanograins that, unexpectedly, presented a smaller surface area for hydrocarbon adsorption and therefore a lower sensitivity. A similar effect was found for tungsten-doped ZnO MS thin films used for the detection of NO_2 traces [27]. Also, in this case, the doped film deposited at a higher pressure had a much lower grain density and porosity and a spiky morphology that rendered a smaller sensitivity than at low pressure.

- Membranes: Porous conductive carbon films are useful for applications in fuel cells and biomedical sensors, and their porosity was investigated using unbalanced MS. Kim et al. [28] found that the porosity of the films and their conductivity highly increased at higher deposition pressures following the typical particle thermalization concepts discussed in Sections 3.1.3 and 3.2.2. Besides thermalization at relatively high pressures, another common approach widely used in the literature and in various technological processes based on membranes is to use patterned substrates to replicate the surface features during film growth, in order to obtain nanostructured films with specific catalytic and separation properties. Examples illustrating this approach will be further discussed in Section 3.3.3.

3.3.2 Nanostructured Magnetron Sputtered Thin Films

The examples discussed in the previous section rely on porous thin films with rather homogenous and small pore sizes. The applied strategy to increase the porosity consisted of increasing the plasma gas pressure during deposition. Herein, we will discuss the use of other deposition strategies to increase the porosity of the thin films and, in addition, generate an open microstructure that, generally formed by separated nanocolumns, can be critical for given applications. We will discuss several methodologies used for the preparation of Li battery electrodes, composite thin-film anodes for fuel cells, and electrodes utilized as electrochemical sensors.

- Thin-film anodes for Li batteries: Silicon is a common electrode for Li ion batteries. The functioning principle of the battery system under negative polarization of the anode entails the reduction of Li^+ cations, the formation of metallic Li and the reversible incorporation of this latter into the silicon anode material, that is:

$$Li^+ + e^- \rightarrow Si(Li)$$

The main problem of these electrodes is the high expansion undergone by the silicon anode (up to 400%) when it incorporates atomic Li. To avoid the collapse of the structure of the silicon electrode, the existence of free space is necessary to accommodate this volume expansion. Besides, the release of Li upon its oxidation when reversibly cycling the system produces a decrease in the electrode volume that, again, must be accommodated by an open microstructure. The literature on the subject refers to different approaches to get open silicon nanostructures able to reversibly accommodate the insertion/release of high amounts of lithium. For example, Mukanova et al. [29] prepared Si anodes by radio frequency (RF) MS deposition of silicon onto a porous Cu current collector. This porous substrate induces extra nanostructuration into the deposited silicon films (see Sections 3.2.6 and 3.3.3), which, in this way, acquires a 3D open nanostructure quite well adapted for reversible volume expansion changes. Examples of reported strategies with the same aim, that is, the accommodation of volume changes in a reversible way, include: the fabrication of C/Si multilayers with a transition thin Ti layer over the copper current collector and an individual layer thickness of about 20 nm [30]; the MS deposition of silicon onto a Ni foam previously coated by graphene prepared by chemical vapor deposition [31]; the formation of Ni/Si core/shell nanosheet arrays by MS deposition of silicon onto a 3D NiO template [32]; or the aforementioned strategy of fabricating porous thin films of silicon by merely increasing the pressure during the MS deposition [33]. This latter work is a neat example of how increasing the working pressure during MS deposition renders very porous thin films where it is possible to accommodate volume expansion modifications upon incorporation/release of lithium. Recently, a critical appraisal of different thin-film deposition methods for the fabrication of silicon anodes has been published [34]. MS-OAD has been also utilized for the fabrication of composite anodes consisting of a carbon-aluminum-silicon multilayer that efficiently accommodates this silicon lattice expansion/retraction found when cycling the anodes [35]. The key point of this nanocolumnar OAD microstructure is the high volume of empty space available for expansion and the gradation of strain due to the different volumetric expansion of the three stacked materials. Thin-film anodes made of materials other than silicon where volumetric constrains are less severe have been also efficiently prepared by MS-OAD [36].

- Fuel cell electrodes: Conventional technologies of preparation of fuel cell electrodes rely on chemical, ink dispersion, doctor blade, and similar procedures aiming at preparing rather thick agglomerated material layers made of powders with a high porosity. The MS technique has been considered a suitable methodology for the preparation of fuel cell electrodes that may comply with the condition of a high surface area and porosity. Brault et al. [37, 38] have extensively used the MS technique coupled with other technologies, such as RF plasmas, for the preparation of electrodes for proton-exchange membrane fuel cells. Generally, these electrodes are made of carbon and contain Pt nanoparticles, which can be deposited either simultaneously or sequentially while adjusting the deposition conditions in order to get porous carbon films with well-dispersed Pt nanoparticles.

  Solid oxide fuel cells (SOFCs) operating at medium and high temperatures convert chemical energy of a fuel such as hydrogen or methane into electricity. As for many other processes involving gas reactions catalyzed by solids, a condition for the efficient operation of these devices (i.e., to achieve a high conversion yield of energy) is that the anode and cathode are porous to offer a good gas management capacity (i.e., easy diffusion in and out of the reaction centers) and a high concentration of active centers accessible for reaction. Anodes are critical fuel cell components in this regard. SOFC anodes for the burning of hydrogen or hydrocarbons consist of a composite material incorporating nanoparticles of an active catalytic phase, usually a metal such as nickel, dispersed in a mixed oxide matrix acting as an ionic conductor of oxygen. Common mixed oxides utilized for this purpose are yttria-stabilized zircona (YSZ) and gadolinia-doped ceria (GDC). The reaction taking place at the electrode adopts the form

$$CH_4 + O^{=} \rightarrow CO_2 + H_2O + e^{-},$$

where $O^{=}$ refers to oxygen ions electrochemically pumped from the solid electrolyte separating anode and cathode and CH_4 is the fuel burned to produce the electricity. Typical anodes for SOFCs are prepared by screen printing or similar procedures and entail rather thick layers of several hundred microns. MS has been proposed as a suitable technology to reduce the amount of expensive material incorporated into the anode.

Maximizing anode efficiency requires various conditions, including a high dispersion of the nickel phase to maximize the number of the three boundary phase sites where the reaction takes place (i.e., contact between the nickel and the oxide) and an open porosity favoring the diffusion of reactants up to the reaction sites and the back-release of products to the reactant stream. Both conditions have been achieved using the MS-OAD technique for the deposition of Ni-YSZ and Ni-GDC and a multilayer of the two composite compounds incorporating little amounts of gold to avoid carbon poisoning of the catalytically active nickel phase [39–41]. According to Fig. 3.17, the nanocolumnar and highly porous amorphous films obtained at room temperature by means of this methodology present a homogeneous and atomically dispersed distribution of the constituent atoms that, upon annealing and reducing at a high temperature, render thicker and rather separated nanocolumns with a good dispersion of Ni crystallites, a high concentration of three boundary points for catalysis, and an optimum topography for easy diffusion of reactants and products to/from the active centers.

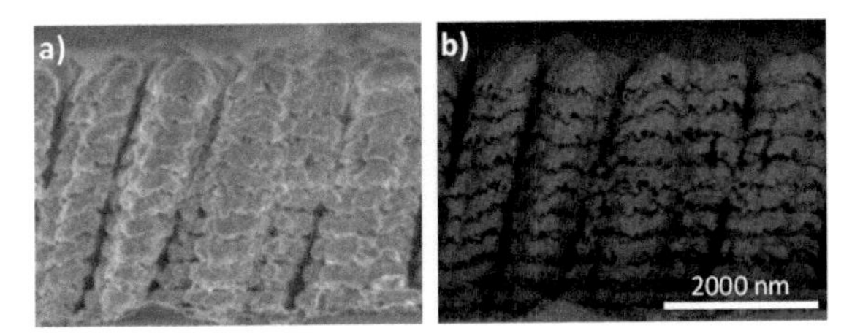

Figure 3.17 Cross-section SEM microgrpahs, corresponding to secondary (a) and backscattered (b) electrons, of Ni-YSZ composite thin films prepared by MS-OAD and used as a thin film anode in SOFCs. The columnar microstructure of these anode films is well suited to favor the diffusion of the reactants up to the reaction sites and the diffusion of the products out of the reaction sites at the thre boundary layer between Ni particles and the YSZ matrix. Reprinted from Ref. [41], Copyright (2017), with permission from Elsevier.

- Liquid phase electrochemical sensors: Electrochemical sensors are widely used for the quantitative analysis of a large variety of components and analytes, either combined with an enzymatic procedure of detection or used as label-free transducers. Electrochemical sensors must be reliable and robust and present a rapid response when they are brought in contact with the analyte of interest. Both sensitivity and rapid response are tightly linked with the existence of a porous and open microstructure favoring the diffusion of the detected molecule from the medium to specific sites at the surface of the electrode. Traditionally, chemical and biochemical detection has been considered within the realm of chemistry and most procedures utilized for the manufacturing of electrodes are based on chemical or electrochemical routes in liquid phase. Recently, MS has been identified as an alternative procedure for the manufacture of sensor electrodes presenting a sensitivity and limit of detection comparable to or even better than those achieved by electrodes prepared through these classical chemical routes. A high-porosity and well-defined microstructure is key for the good performance of electrodes deposited by PVD procedures. In addition, these electrodes may also present other advantages with regard to those prepared through classical chemical routes, such as the straightforward synthesis of composite and multilayer electrodes and the fact that electrode films can be manufactured at room temperature and, therefore, are compatible with any kind of substrate. These advantages have been highlighted recently in a review paper on the use of oxide thin films prepared by MS for pH sensor fabrication [42]. Different MS deposition approaches, electrode compositions, and nanostructures have been explored in this field. For example, deposition conditions rendering porous thin films or leading to the fabrication of well-dispersed metal nanoparticles on columnar or nanostructured films have been utilized for the fabrication of electrochemical sensors for the detection of dopamine [43]. Mixed composition porous electrodes for the non-enzymatic detection of glucose have been also prepared upon adjustment of deposition conditions [44–46]. In this topic, the use of MS-OAD has clearly exemplified the advantages of this modified deposition configuration over classical MS procedures for the preparation of nanostructured, highly porous thin films. For example, zigzag nanocolumnar Ti-Ag thin films have been prepared by this methodology, showing that the thin-film architecture (i.e., the number of zigzag periods or separation among columns due to variations in the deposition angle) has a pronounced effect on the sensor response of the system. In our group, we have also shown that WO_3 thin films prepared by MS-OAD present higher stability and a wider linear response range when used for pH sensing applications [47]. A similar result was evidenced for NiO thin films prepared by MS-OAD and utilized for the non-enzymatic determination of glucose [48]. Moreover, the high-porosity and nanocolumnar microstructure of OAD thin films has demonstrated to be quite efficient for the development of enzymatic-based electrodes where the incorporation and attachment of enzymes to an inorganic substrate is a key issue for an adequate system response. This possibility has been recently demonstrated by our group with the development of a cholesterol sensor electrode formed by a nanocolumnar Pt electrode (see Fig. 3.18 for a more precise observation of the microstructure of the electrodes) prepared by MS-OAD and a suitable enzyme incorporated and attached to its surface [49].

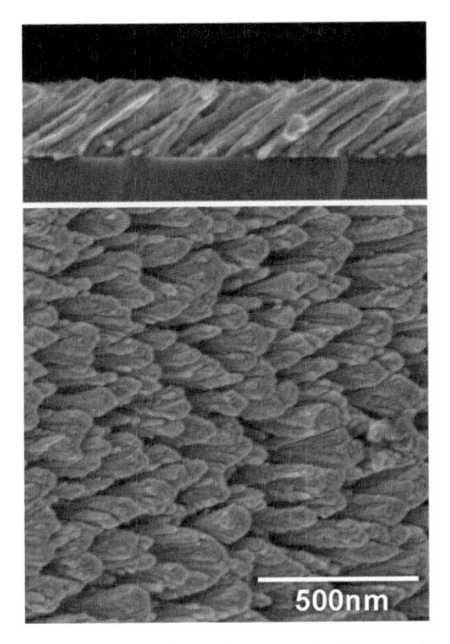

Figure 3.18 Cross-section (top) and normal (bottom) SEM micrographs of Pt films prepared by MS-OAD and used as an electrode for cholesterol determination (Ref. [49]).

3.3.3 Nanostructured Thin Films Deposited on Patterned or Nanostructured Substrates

Another possibility for the fabrication of nanostructured thin films for a large variety of applications is the MS deposition onto substrates with a nano- or microstructured surface. A general description of the influence of the substrate on MS deposition has been presented in Section 3.2.6. Effectively, in a recent work using an OAD configuration, we have shown that the surface state of a substrate strongly affects the growth regime of a thin film deposited by MS onto its surface [17]. In particular, considering a surface with a periodic rippled structure, we have shown that the wavelength defined by the separation between ripples rather than by their height is the main factor influencing the type of microstructure and final roughness of the films deposited by MS. Moreover, we have been able to show that not only the microstructure but also the chemistry of SiO_x thin films can be patterned laterally when deposited by reactive MS-OAD onto a patterned substrate [50]. Figure 3.19 illustrates the possibilities of this methodology for SiO_x/SiO_2 nanostructured surfaces where the composition follows the variation defined by the linear topology of the patterns prepared by laser interference.

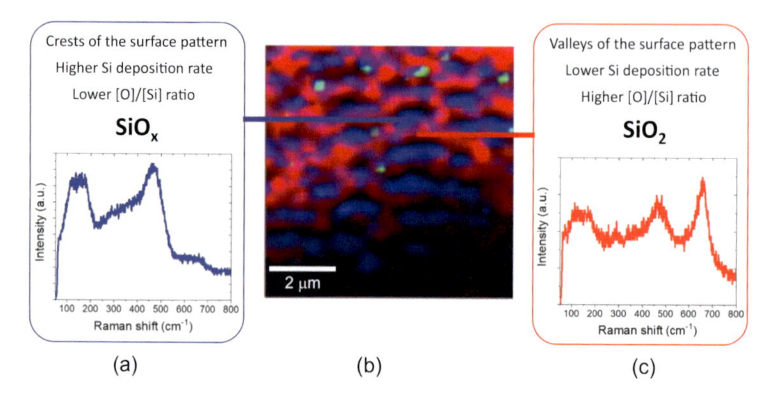

Figure 3.19 (b) Raman-AFM compositional map of a SiO_x/SiO_2 layer prepared by MS-OAD on a patterned substrate. (a, c) The panels present the Raman spectra recorded for the SiO_x ($x < 2$) and SiO_2 zones. Reprinted from Ref. [51], Copyright (2019), with permission from Elsevier.

Previous principles are new, and in the absence of a systematic framework to account for the nanostructuration phenomena induced by substrate features, most practical developments using patterned substrates for the MS fabrication of nanostructure thin films have relied on an empirical basis. This has been used in fields such as sensors, fuel cells, and, most commonly, membranes for fluid separation. Porous alumina and porous silicon, with pore sizes in the range of tenths of nanometers have been used as common substrates for these applications under the leading concept that the deposited thin film replicates the pattern array defined at the surface of the membranes. The main advantage of the use of these membranes as a template for thin-film nanostructuring is their easy availability and their fabrication by well-established electrochemical methods. Numerous examples can be found in the literature utilizing this approach, including the use of porous silicon substrates for the deposition of WO_3 and ZnO thin films used as sensors of ammonia and other gases [43, 51], and numerous examples of different materials deposited onto porous alumina membranes. Examples of this latter possibility encompass the deposition of YSZ for fuel cell applications [52]; nanoporous silicon dioxide membranes [53]; ZnO and SnO_2 acting as membranes, the latter in the form of nanobaskets [54, 55]; metals such as Ni [56, 57] or Pd-Cu and Pd-Ag alloys for hydrogen separation [58, 59]; oxides such as NiO [60]; nitrides such as nanostructured TiN in the form of hollow structures, nanotubes, and nanocages [61]; or even the transparent and conducting indium tin oxide [62].

Besides these common substrates, particularly for separation and filtration membranes, there are numerous examples using other patterned substrates to control the nanostructure of the thin films deposited by MS. For example, nanostructured metal thin films have been deposited onto nanostructured polymers for gas separation purposes [63–65] or as Pt electrodes to promote the oxygen reduction reaction in proton-exchanged fuel cells [56]. Examples of other patterned substrates are the use of carbon nanotubes for the growth of Pt or Pd membranes onto a patterned substrate consisting of porous stainless steel covered with YSZ film deposited by spray pyrolysis [66].

These examples and many others reported in the literature clearly support the wide range of possibilities offered by this MS approach in many technologically relevant fields, as well as the need to systematize the deposition procedures for precise control of the nanostructure of the deposited layers.

References

1. D. M. Mattox. *Handbook of Physical Vapor Deposition (PVD) Processing*, 2nd ed., William Andrew (2010).
2. K. Sarakinos, J. Alami, S. Konstantinidis, High power pulsed magnetron sputtering: A review on scientific and engineering state of the art, *Surf. Coat. Technol.*, **204** (2010) 1661.
3. T. Nakano, Y. Saitou, K. Oya, Transient evolution of the target erosion profile during magnetron sputtering: dependence on gas pressure and magnetic configuration, *Surf. Coat. Technol.*, **326** (2017) 436.
4. J. A. Thornton, Influence of apparatus geometry and deposition conditions on the structure and topography of thick sputtered coatings, *J. Vac. Sci. Technol.*, **11** (1974) 666.
5. A. Barranco, A. Borrás, A. R. González-Elipe, A. Palmero, Perspectives on oblique angle deposition of thin films: from fundamentals to devices, *Prog. Mater. Sci.*, **78** (2016) 59.

6. D. Depla, S. Mahieu (eds.), *Reactive Sputter Deposition*, Springer Series in Materials Science, Springer-Verlag, Berlin, Heildelberg (2008).
7. A. Palmero, E. D. van Hattum, W. M. Arnoldbik, A. M. Vredenberg, F. H. P. M. Habraken, Characterization of the plasma in a radio-frequency magnetron sputtering system, *J. Appl. Phys.*, **95** (2004) 7611.
8. A. Palmero, H. Rudolph, F. H. P. M. Habraken, One-dimensional analysis of the rate of plasma-assisted sputter deposition, *J. Appl. Phys.*, **101** (2007) 083307.
9. A. Palmero, H. Rudolph, F. H. P. M. Habraken, Generalized Keller-Simmons formula for nonisothermal plasma-assisted sputtering depositions, *Appl. Phys. Lett.*, **89** (2006) 211501.
10. R. Alvarez, A. Garcia-Valenzuela, C. Lopez-Santos, F. J. Ferrer, V. Rico, E. Guillen, M. Alcon-Camas, R. Escobar-Galindo, A. R. Gonzalez-Elipe, A. Palmero, High-rate deposition of stoichiometric compounds by reactive magnetron sputtering at oblique angles, *Plasma Processes Polym.*, **13** (2016) 960.
11. A. Siad, A. Besnard, C. Nouveau, P. Jacquet, Critical angles in DC magnetron glad thin films, *Vacuum*, **131** (2016) 305.
12. J. A. Thornton, Influence of substrate temperature and deposition rate on structure of thick sputtered Cu coatings, *J. Vac. Sci. Technol.*, **12** (1975) 830.
13. R. Alvarez, J. M. Garcia-Martin, A. Garcia-Valenzuela, M. Macias-Montero, F. J. Ferrer, J. Santiso, V. Rico, J. Cotrino, A. R. Gonzalez-Elipe, A. Palmero, Growth regimes of porous gold thin films deposited by magnetron sputtering at oblique incidence: from compact to columnar microstructures, *Nanotechnology*, **24** (2013) 045604.
14. R. Alvarez, J. M. Garcia-Martin, A. Garcia-Valenzuela, M. Macias-Montero, F. J. Ferrer, J. Santiso, V. Rico, J. Cotrino, A. R. Gonzalez-Elipe, A. Palmero, Nanostructured Ti thin films by magnetron sputtering at oblique angles, *J. Phys. D: Appl. Phys.*, **49** (2016) 045303.
15. R. Alvarez, C. Lopez-Santos, F. J. Ferrer, V. Rico, J. Cotrino, A. R. Gonzalez-Elipe, A. Palmero, Modulating low energy ion plasma fluxes for the growth of nanoporous thin films, *Plasma Processes Polym.*, **12** (2015) 719.
16. M. Macias-Montero, F. J. Garcia-Garcia, R. Alvarez, J. Gil-Rostra, J. C. Gonzalez, J. Cotrino, A. R. Gonzalez-Elipe, A. Palmero, Influence of plasma-generated negative oxygen ion impingement on magnetron sputtered amorphous SiO2 thin films during growth at low temperatures, *J. Appl. Phys.*, **111** (2012) 054312.

17. S. Gauter, F. Haase, H. Kersten, Experimentally unraveling the energy flux originating from a DC magnetron sputtering source, *Thin Solid Films*, **669** (2019) 8.
18. A. Garcia-Valenzuela, S. Muñoz-Piña, G. Alcala, R. Alvarez, B. Lacroix, A. J. Santos, J. Cuevas-Maraver, V. Rico, R. Gago, L. Vazquez, J. Cotrino, A. R. Gonzalez-Elipe, A. Palmero, Growth of nanocolumnar thin films on patterned substrates at oblique angles, *Plasma Process. Polym.*, **16** (2019) 1800135.
19. R. Alvarez, R. Romero-Gómez, J. Gil-Rostra, J. Cotrino, F. Yubero, A. Palmero, A. R. González-Elipe, On the microstructure of thin films grown by an isotropically directed deposition flux, *J. Appl. Phys.*, **108** (2010) 064316.
20. S. J. Gregg, K. S. W. Sing, *Adsorption, Surface Area and Porosity*, Academic Press, London (1982).
21. R.-T. Wen, C. G. Granqvist, G. A. Niklasson, Eliminating degradation and uncovering ion-trapping dynamics in electrochromic WO_3 thin films, *Nat. Mater.*, **14** (2015), 996.
22. X. Sun, Z. Liu, H. Cao, Effects of film density on electrochromic tungsten oxide thin films deposited by reactive dc-pulsed magnetron sputtering, *J. Alloys Compd.*, **504** (2010) S418.
23. C. Salinga, O. Kappertz, M. Wuttig, Reactive direct current magnetron sputtering of tungsten oxide: A correlation between film properties and deposition pressure, *Thin Solid Films*, **515** (2006) 2760.
24. J. Gil-Rostra, M. Cano, J. M. Pedrosa, F. J. Ferrer, F. García-García, F. Yubero, A. R. González-Elipe, Electrochromic behavior of $W_xSi_yO_z$ thin films prepared by reactive magnetron sputtering at normal and glancing angles, *ACS Appl. Mater. Interfaces*, **22** (2012) 628.
25. Y. Shen, B. Zhang, X. Cao, D. Wei, J. Ma, Jia, S. Gao, B. Cui, Y. Jin, Microstructure and enhanced H2S sensing properties of Pt-loaded WO_3 thin films, *Sens. Actuators, B*, **193** (2014) 273.
26. S. Fairose, S. Ernest, S. Daniel, Effect of oxygen sputter pressure on the structural, morphological and optical properties of ZnO thin films for gas sensing application, *Sens. Imaging*, **19** (2018) 1.
27. M. Kwoka, B. Lyson-Sypien, A. Kulis, M. Maslyk, M. A. Borysiewicz, E. Kaminska, J. Szuber, Surface properties of nanostructured, porous ZnO thin films prepared by direct current reactive magnetron sputtering, *Materials*, **11** (2018) 131.
28. T. Tesfamichael, C. Cetin, C. Piloto, M. Arita, J. Bell, The effect of pressure and W-doping on the properties of ZnO thin films for NO_2 gas sensing, *Appl. Surf. Sci.*, **357** (2015) 728.

29. S. I. Kim, B. B. Sahu, B. M. Weon, J. G. Han, J. Koskin, S. Franssila, Making porous conductive carbon films with unbalanced magnetron sputtering, *Jpn. J. Appl. Phys.*, **54** (2015) 010304.
30. A. Mukanova, A. Nurpeissova, S.-S. Kim, M. Myronov, Z. Bakenov, N-type doped silicon thin film on a porous Cu current collector as the negative electrode for Li-ion batteries, *ChemistryOpen*, **7** (2018) 92.
31. J. Wang, S. Li, Y. Zhao, J. Shi, L. Lv, H. Wang, Z. Zhang, W. Feng, The influence of different Si : C ratios on the electrochemical performance of silicon/carbon layered film anodes for lithium-ion batteries, *RSC Adv.*, **8** (2018) 6660.
32. A. Mukanova, A. Nurpeissova, A. Urazbayev, S.-S. Kim, M. Myronov, Z. Bakenov, Silicon thin film on graphene coated nickel foam as an anode for Li-ion batteries, *Electrochim. Acta*, **258** (2017) 800.
33. X. H. Huang, P. Zhang, J. B. Wu, Y. Lin, R. Q. Guo, Nickel/silicon core/shell nanosheet arrays as electrode materials for lithium ion batteries, *Mater. Res. Bull.*, **80** (2016) 30.
34. M. T. Demirkan, L. Trahey, T. Karabacak, , Low-density silicon thin films for lithium-ion battery anodes, *Thin Solid Films*, **600** (2016) 126.
35. A. Mukanova, A. Jetybayeva, S.-T. Myung, S.-S. Kim, Z. Bakenov, A mini-review on the development of Si-based thin film anodes for Li-ion batteries, *Mater. Today Energy*, **9** (2018) 49.
36. R. Krishnan, M. Lu, N. Koratkar, Functionally strain-graded nanoscoops for high power Li-ion battery anodes, *Nano Lett.*, **11** (2011) 377.
37. A-H. Zinn, Combinatorial investigation of solid-state-battery materials, PhD thesis, Fakultät für Maschinenbau der Ruhr-Universität Bochum (2016).
38. H. Rabat, P. Brault, Plasma sputtering deposition of PEMFC porous carbon platinum electrodes, *Fuel Cells*, **2** (2008) 81.
39. M. Mougenot, P. Andreazza, C. Andreazza-Vignolle, R. Escalier, Th. Sauvage, O. Lyon, P. Brault, Cluster organization in co-sputtered platinum-carbon films as revealed by grazing incidence X-ray scattering, *J. Nanopart. Res.*, **14** (2012) 672.
40. F. J. García-García, F. Yubero, J. P. Espinós, A. R. González-Elipe, R. M. Lambert, Synthesis, characterization and performance of robust poison-resistant ultrathin film yttria stabilized zirconia-nickel anodes for application in solid electrolyte fuel cells, *J. Power Sources*, **324** (2016) 679.
41. F. J. García-García, A. M. Beltran, F. Yubero, A. R. González-Elipe, R. M. Lambert, High performance novel gadolinium doped ceria/yttria

stabilized zirconia/nickel layered and hybrid thin film anodes for application in solid oxide fuel cells, *J. Power Sources*, **363** (2017) 251.

42. F. J. García-García, F. Yubero, A. R. González-Elipe, R. M. Lambert, Microstructural engineering and use of efficient poison resistant Au-doped Ni-GDC ultrathin anodes in methane-fed solid oxide fuel cells, *Int. J. Hydrogen Energy*, **43** (2018) 885.
43. D. K. Maurya, A. Sardarinejad, K. Alameh, Recent developments in R.F. magnetron sputtered thin films for pH sensing applications: an overview, *Coatings*, **4** (2014) 756.
44. X. Mei, Q. Wei, H. Long, Z. Yu, Z. Deng, L. Meng, J. Wang, J. Luo, C.-T. Lin, L. Ma, K. Zheng, N. Hu, Long-term stability of Au nanoparticle-anchored porous boron-doped diamond hybrid electrode for enhanced dopamine detection, *Electrochim. Acta*, **271** (2018) 84.
45. C.-W. Liu, W. E. Chen, Y. T. A. Sun, C.-R. Lin, Fabrication and electrochemistry characteristics of nickel-doped diamond-like carbon film toward applications in non-enzymatic glucose detection, *Appl. Surf. Sci.*, **436** (2018) 967.
46. D. Jung, R. Ahmad, Y. B. Hahn, Nonenzymatic flexible field-effect transistor based glucose sensor fabricated using NiO quantum dots modified ZnO nanorods, *J. Colloid Interface Sci.*, **512** (2018) 21.
47. G. Ynag, E. Liu, N. W. Khun, S. P. Jiang, Direct electrochemical response of glucose at nickel-doped diamond like carbon thin film electrodes, *J. Electroanal. Chem.*, **627** (2009) 51.
48. P. Salazar, F. J. García-García, F. Yubero, J. Gil-Rostra, A. R. González-Elipe, Characterization and application of a new pH sensor based on magnetron sputtered porous WO_3 thin films deposited at oblique angles, *Electrochim. Acta*, **193** (2016) 24.
49. F. J. García-García, P. Salazar, F. Yubero, A. R. Gonzalez-Elipe, Non-enzymataic glucose electrochemical sensor made of porous NiO thin films prepared by reactive magnetron sputtering at oblique angles, *Electrochim. Acta*, **201** (2016) 36.
50. M. Martín, P. Salazar, R. Alvarez, A. Palmero, C. López-Santos, J. L. González-Mora, A. R. González-Elipe, Cholesterol biosensing with a polydopamine-modified nanostructured platinum electrode prepared by oblique angle physical vacuum deposition, *Sens. Actuators, B*, **240** (2017) 37.
51. A. García-Valenzuela, R. Alvarez, V. Rico, J. P. Espinós, M. C. López-santos, J. Solís, J. Siegel, A. del Campo, A. Palmero, A. R. González-Elipe, 2D compositional self-patterning in magnetron sputtered thin films, *Appl. Surf. Sci.*, **480** (2019) 115.

52. H. S. Al-Salman, M. J. Abdullah, Preparation of ZnO nanostructures by RF-magnetron sputtering on thermally oxidized porous silicon substrate for VOC sensing application, *Measurement*, **59** (2015) 248.
53. Y. Lim, S. Hong, J. Bae, H. Yang, Y. B. Kim, Influence of deposition temperature on the microstructure of thin-film electrolyte for SOFCs with a nanoporous AAO support structure, *Int. J. Hydrogen Energy*, **42** (2017) 10199.
54. F. Chen, A. H. Kitai, Influence of deposition temperature on the microstructure of thin-film electrolyte for SOFCs with a nanoporous AAO support structure, *Thin Solid Films*, **517** (2008) 622.
55. P. L. Johnson, D. Teeters, Formation and characterization of SnO_2 nanobaskets, *Solid State Ionics*, **177** (2006) 2821.
56. G. Q. Ding, W. Z. Shen, M. J. Zheng, D. H. Fan, Synthesis of ordered large-scale ZnO nanopore arrays, *Appl. Phys. Lett.*, **88** (2006) 103106-3.
57. G. Sievers, T. Vidakovic-Koch, C. Walter, F. Steffen, S. Jakubith, A. Krut, Ultra-low loading Pt-sputtered gas diffusion electrodes for oxygen reduction reaction, *J. Appl. Electrochem.*, **48** (2018) 221.
58. V. Perekrestov, A. Kornyushchenko, V. Natalich, S. Ostendorp, G. Wilde, Formation of porous nickel nanosystems using alumina membranes as templates for deposition, *Mater. Lett.*, **153** (2015) 171.
59. A. I. Pereira, P. Pérez, S. C. Rodrigues, A. Mendes, L. M. Madeira, C. J. Tavares, Deposition of Pd–Ag thin film membranes on ceramic supports for hydrogen purification/separation, *Mater. Res. Bull.*, **61** (2015) 528.
60. S.-K. Ryi, J.-S. Park, K.-R. Hwang, D.-W. Kim, H.-S. An, Pd-Cu alloy membrane deposited on alumina modified porous nickel support (PNS) for hydrogen separation at high pressure, *Korean J. Chem. Eng.*, **29** (2012) 59.
61. A. Gutiérrez-Delgado, G. Domínguez-Cañizares, J. A. Jiménez, I. Preda, D. Díaz-Fernández, F. Jiménez-Villacorta, G. R. Castro, J. Chaboy, L. Soriano, Hexagonally-arranged-nanoporous and continuous NiO films with varying electrical conductivity, *Appl. Surf. Sci.*, **276** (2013) 832.
62. O. Sánchez, M. Hernández-Vélez, D. Navasa, M. A. Auger, J. L. Baldonedo, R. Sanza, K. R. Pirota, M. Vázquez, Functional nanostructured titanium nitride films obtained by sputtering magnetron, *Thin Solid Films*, **495** (2006) 149.
63. G. Q. Ding, W. Z. Shen, M. J. Zheng, Z. B. Zhou, Integration of single-crystalline nanocolumns into highly ordered nanopore arrays, *Nanotechnology*, **17** (2006) 2590.

64. M. J. Detisch, T. J. Balk, D. Bhattacharyy, Synthesis of catalytic nanoporous metallic thin films on polymer membranes, *Ind. Eng. Chem. Res.*, **57** (2018) 4420.

65. M. Shaban, A. M. Ahmed, E. Abdel-Rahman, H. Hamdy, Morphological and optical properties of ultra-thin nanostructured Cu films deposited by RF sputtering on nanoporous anodic alumina substrate, *Micro Nano Lett.*, **11** (2016) 295.

66. S. Cuynet, A. Caillard, S. Kaya-Boussougou, T. Lecas, N. Semmar, J. Bigarre, P. Buvat, P. Brault, Membrane patterned by pulsed laser micromachining for proton exchange membrane fuel cell with sputtered ultra-low catalyst loadings, *J. Power Sources*, **298** (2015) 299.

67. Y. Huang, R. Dittmeyer, Preparation of thin palladium membranes on a porous support with rough surface, *J. Membr. Sci.*, **302** (2007) 160.

Chapter 4

Atomic Species Generation by Plasmas

Rok Zaplotnik, Gregor Primc, Domen Paul, Miran Mozetič, Janez Kovač, and Alenka Vesel
Department of Surface Engineering, Jozef Stefan Institute, Jamova Cesta 39, 1000 Ljubljana, Slovenia
rok.zaplotnik@ijs.si; gregor.primc@ijs.si; domen.paul@ijs.si; miran.mozetic@ijs.si; janez.kovac@ijs.si; alenka.vesel@ijs.si

4.1 Introduction

Plasmas of molecular gases like oxygen; nitrogen; hydrogen; ammonia; sulfur dioxide; hydrogen sulfide; water vapor; hydrocarbons; fluorinated, nitrated, and oxidized hydrocarbons; and carbon dioxide are widely used for tailoring surface properties of solid materials. The molecules are partially ionized, dissociated, and/or excited, because of the plasma conditions. The plasma created in a molecular gas is, therefore, rich in molecular fragments, which interact with each other and with the solid materials facing the plasma. The molecular fragments are often called "radicals" and may be atoms, diatomic molecules, or larger fragments of original molecules. The major difference between the original molecules and their fragments is in chemical reactivity. The original molecules are

Plasma Applications for Material Modification: From Microelectronics to Biological Materials
Edited by Francisco L. Tabarés

ISBN 978-981-4877-35-0 (Hardcover), 978-1-003-11920-3 (eBook)
www.jennystanford.com

rather stable, and they will usually not interact with solid materials at low temperatures. The fragments are chemically unstable and tend to interact with any object placed in the gaseous plasma, including themselves. An illustrative example is methane (CH_4). The original molecule CH_4 is regarded as inert at room temperature. Upon colliding with a solid material the molecule will just bounce off from the surface. Because of plasma conditions, however, the CH_4 molecule is dissociated into fragments, such as CH_3, CH_2, CH, C, and H. The radicals will interact with any object placed in the gaseous plasma. The interaction of carbon atoms with a solid material causes condensation of the carbon atom on the surface of the solid material. A carbon atom arriving at a solid material will stick to the surface with a very high probability, almost 100%. The sticking coefficient depends on the temperature of the solid material but is regarded to be very high even for the highest temperatures available for solid materials before they melt or sublimate. The sticking coefficient of CH_x radicals decreases with the increasing content of hydrogen. It is very high for CH and marginal for CH_4 [1].

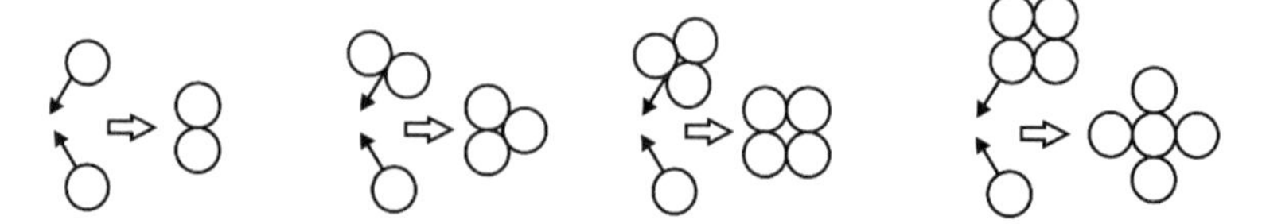

Figure 4.1 Schematic of cluster formation in a gaseous plasma.

The interaction of radicals with each other often leads to the formation of molecules in a gaseous plasma. Such an effect is shown schematically and in a rather simplified manner in Fig. 4.1. The molecules formed on the interaction of gaseous radicals may or may not be stable. Oxygen, for example, forms a stable two- or three-atom molecule: O_2 and O_3 (ozone). Carbon, on the other hand, forms clusters of an almost arbitrary number of atoms. The association of the atoms in the gas phase as shown schematically in Fig. 4.1 is governed by the natural rules, in particular, conservation of energy and momentum. By definition, a molecule in the ground state will have the potential energy 0. When the molecule dissociates into atoms, the potential energy of the atoms is equal to the binding energy of the parent molecule. The dissociation energies of simple molecules are shown in Table 4.1. The dissociation energy is on the order of a few electron volts. This energy should be preserved upon the association of the atoms into a molecule. The excessive energy before the association could be transformed either into the internal energy of the formed molecule or into its kinetic energy. If all the potential energy of atoms before a collision is transformed into the internal energy of the formed molecule, the molecule would dissociate immediately because the internal energy of the formed molecule would be equal to the dissociation energy.

Table 4.1 Dissociation energies of selected molecules

Molecule	Dissociation energy
O_2	5.16 eV
N_2	9.8 eV
H_2	4.52 eV

Source: [2]

The conservation of momentum, however, forbids the transformation of the potential energy before a collision into the kinetic energy of the formed molecule. The simple reactions shown in Fig. 4.2 are, therefore, forbidden because of the requirement of the conservation of energy and momentum. The consequence of this simple explanation is perfect stability of radicals, such as neutral atoms, under vacuum conditions. A possible way for conserving energy and momentum in simple association reactions illustrated in Fig. 4.2 is radiative association. In this case, association occurs simultaneously with the formation of a light quantum (a photon), which holds the available energy and momentum. Such collisions are unlikely to occur in examples shown in Fig. 4.2 but may be predominant in the plasmas created in gas mixtures. A good example is the association of an O atom with a N atom, which results in a highly excited NO molecule that de-excites readily by radiation in the UV range of the spectrum [3].

$$\left.\begin{array}{l} O+O\rightarrow O_2 \\ N+N\rightarrow N_2 \\ H+H\rightarrow H_2 \end{array}\right\} \text{These reactions are forbidden.}$$

Figure 4.2 Some forbidden association reactions between atoms in the gas phase.

If the density of gaseous particles is high enough there is a possibility that three particles will appear in a small volume where the distance between the particles is small enough to allow them to interact. This condition will lead to a three-body collision. Some simple thee-body collisions are shown in Fig. 4.3.

$$O + O + O_2 \rightarrow O_2 + O_2$$
$$N + N + N_2 \rightarrow N_2 + N_2$$
$$H + H + H_2 \rightarrow H_2 + H_2$$

Figure 4.3 Some reactions allowed in the gas phase.

The probability of a three-body collision obviously depends on the density of the gaseous particles, which, in turn, depends on the pressure

$$p = nkT, \tag{4.1}$$

where p is the gas pressure, n is the density of particles, k is Boltzmann constant (k = 1.38064852 × 10^{-23} m^2 kg s^{-2} K^{-1}), and T is the gas temperature expressed in Kelvin (K). The probability for three-body collisions is often expressed in terms of the collision frequency (ν), which was found to be roughly

$$\upsilon = \sigma^{5/2} n^2 \langle v \rangle, \tag{4.2}$$

where σ is the collisional cross section, n is the density of the particles, and $\langle v \rangle$ is the average speed of the random (thermal) motion of the particles, that is

$$\langle v \rangle = \sqrt{\frac{8kT}{\pi m}}. \tag{4.3}$$

Here, k is the Boltzmann constant, T is the gas temperature, and m is the mass of a particle. The relation shown in Eq. 4.2 is a rough estimation obtained by rounding some constants and taking into account a rigid-sphere model. In the plasma created in a simple molecular gas, the collision frequency of three-body collisions is roughly 10^7 s^{-1} (ten million collisions per second only) at atmospheric pressure and room temperature. The relation in Eq. 4.2 shows that the three-body collision depends on the square of the pressure as long as the temperature is almost constant. Rough estimates of the three-body collision frequency at various pressures and room temperature are indicated in Table 4.2.

Table 4.2 Rough estimates of the three-body collision frequency at different pressures

Pressure (Pa)	Pressure (mbar)	Pressure (torr)	Pressure (bar)	3-body collision frequency (s^{-1})	Atom lifetime
0.1	10^{-3}	7.5×10^{-4}	10^{-6}	10^{-5}	30 h
0.3	3×10^{-3}	2.3×10^{-3}	3×10^{-6}	10^{-4}	3 h
1	10^{-2}	7.5×10^{-3}	10^{-5}	10^{-3}	20 min
3	3×10^{-2}	2.3×10^{-2}	3×10^{-5}	10^{-2}	2 min
10	10^{-1}	7.5×10^{-2}	10^{-4}	10^{-1}	10 s
30	3×10^{-1}	2.3×10^{-1}	3×10^{-4}	1	1 s
100	1	7.5×10^{-1}	10^{-3}	10^{1}	0.1 s
300	3	2.3	3×10^{-3}	10^{2}	10 ms
1000	10	7.5	10^{-2}	10^{3}	1 ms
3000	30	2.3×10^{1}	3×10^{-2}	10^{4}	0.1 ms
10,000	10^{2}	7.5×10^{1}	10^{-1}	10^{5}	10 μs
30,000	3×10^{2}	2.3×10^{2}	3×10^{-1}	10^{6}	1 μs
100,000	10^{3}	7.5×10^{2}	1	10^{7}	0.1 μs

A three-body collision, however, does not ensure association of two atoms into the parent molecule. If all atoms that have suffered a three-body collision actually associate into parent molecules, as shown in Fig. 4.3, the lifetime of the atoms would be inversely proportional to the collision frequency. The lifetime in this rough (not fully justified) approximation is added to Table 4.2, just for a rough guidance.

The neutral atoms may or may not associate to parent molecules according to the three-body collisions presented in Fig. 4.3. If the probability of such homogeneous reactions at the three-body collisions would be close to 1, then the lifetime of an atom would be as shown in Table 4.2. The observed lifetimes are somewhat larger; therefore, the values presented in Table 4.2 are just to get an impression regarding the order of magnitude. The lifetimes at pressures below a few Pascals are really large and have never been reported for laboratory experiments because other effects leading to the loss of radicals prevail.

If the pressure is low, the loss of atoms by a three-body collision is marginal. Vacuum systems suitable for the sustenance of a non-equilibrium gas of a high atom density are usually pumped continuously. The reaction chambers are often cylindrical. They are pumped into one side, whereas on the other side, a molecular gas is introduced via a needle valve or a flow controller. A schematic of the simplest experimental chamber is shown in Fig. 4.4. The gas flows from the right to the left because of the pressure gradient. The pressure at the valve inlet is often atmospheric. The pressure drops along the narrow tube between the valve and the vacuum chamber. If the diameter of the narrow tube is much smaller than that of the vacuum chamber and the pumping speed of the vacuum pump is much larger than the conductance of the narrow tube, the major pressure drop is between the valve and the exhaust of the narrow tube to the vacuum chamber, as shown in Fig. 4.4. The gas drift velocity along the narrow tube and at the exhaust is close to the maximal possible value, which is equal to the sound velocity in gas. The sound velocity in air at room temperature is approximately 343 m/s. The gas expanding into the vacuum chamber forms a jet, as shown in Fig. 4.4. Far away from the narrow tube, the gas drift assumes the normal radial velocity distribution—the drift velocity is the largest at the axis of the tube and the lowest at the edges because of the finite viscosity of the gas. The average drift velocity of the gas in the tube depends on the effective pumping speed according to

$$v_{\mathrm{d}} = S/A, \tag{4.4}$$

where v_{d} is the gas drift velocity as indicated in Fig. 4.4, S is the effective pumping speed, and A is the cross section of the vacuum chamber. In many practical cases, the conductivity of the vacuum chamber for gas flow is much larger than the pumping speed of the vacuum pump. Therefore, one can take into account the pump's pumping speed as S in Eq. 4.4 and calculate the drift velocity along the vacuum tube. This approximation is often justified at an elevated pressure, because the conductivity of vacuum elements increases with increasing pressure. If the condition is fulfilled, the pressure gradient along the vacuum tube will be marginal.

Figure 4.5 represents typical values for the gas drifting through a vacuum chamber. The chamber is assumed to be cylindrical; thus the cross section A (Eq. 4.4) is $\pi d^2/4$, where d is the tube diameter. The gas in the vacuum tube spends some time, which is often referred as the residence time. In a good approximation, the residence time of gas in a simple vacuum chamber, as in Fig. 4.4, is L/v_d, where L is the length of the vacuum tube. Typical residence times for a vacuum tube of length $L = 1$ m are plotted in Fig. 4.6 versus the tube diameter for selected pumping speeds of the vacuum pump.

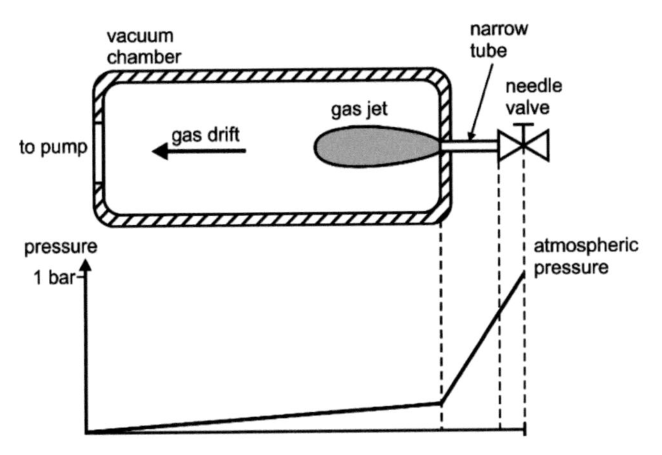

Figure 4.4 Schematic of a simple experimental configuration.

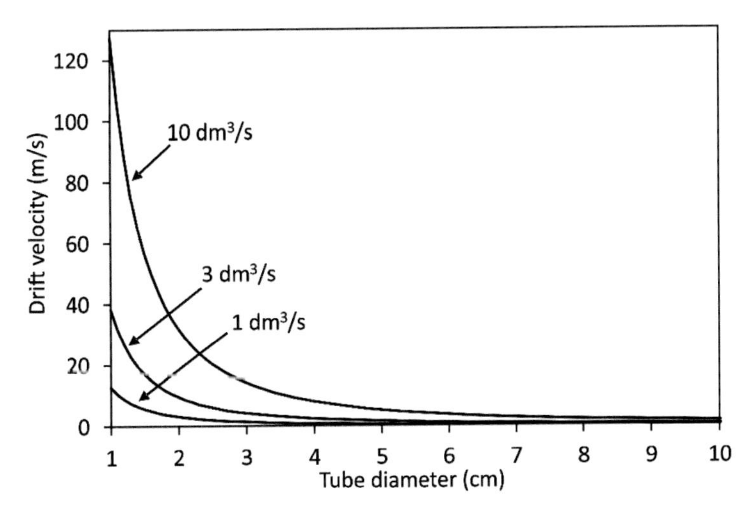

Figure 4.5 Rough estimation of the gas drift velocity versus the vacuum tube diameter for pumping speeds of 1, 3, and 10 dm^3/s.

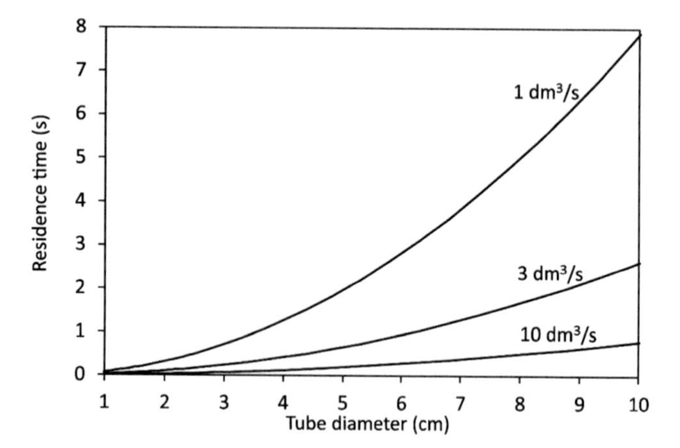

Figure 4.6 Typical residence times for a vacuum tube of length $L = 1$ m versus the tube diameter for selected pumping speeds of the vacuum pump.

The values in Fig. 4.6 give a rough estimation of the gas residence time, that is, the time the gas spends inside the vacuum chamber shown schematically in Fig. 4.4 before it is pumped away. Although they were obtained using rough assumptions, they represent valuable estimates. For typical systems, the residence time is between 1 ms and 1 s. The gaseous radicals that are formed in such a vacuum chamber because of the plasma conditions are therefore pumped away much faster than lost by a three-body collision in the gas phase (see the column titled "Atom lifetime" in Table 4.2). Any discussion about the loss of atoms in the gas phase at pressures below, say, 10 Pa is, therefore, unnecessary.

The atoms may recombine with parent molecules also on surfaces. The flux of atoms onto a surface is

$$j_{\text{atoms}} = \tfrac{1}{4} n \langle v \rangle . \tag{4.5}$$

In the rigid sphere approximation, the atoms are just reflected from the solid surface; thus the kinetic energy is conserved and only the normal component of the velocity becomes inverse. The atoms, however, are chemically reactive and thus tend to interact with the solid material, and the rigid sphere approximation is usually not justified. Because the atoms feel an attractive force, they will stay on the surface for a certain amount of time, which depends on the surface properties. The residence time is often large enough for another atom from the gas phase to arrive to the binding side. The atom arriving from the gas phase interacts strongly with an atom already adsorbed on the surface, and association with the parent molecule occurs. The molecule formed on the surface by the association of two atoms leaves the surface almost immediately. The association depends on the surface density of the atoms and the strength of the binding sites on the surface. The probability of surface association is usually expressed in terms of the recombination coefficient γ, which is defined as a ratio between the flux of atoms to the surface and flux of molecules (which have been formed by surface association of atoms) from the surface. For two-atom molecules, the recombination coefficient is

$$\gamma = \frac{j_{\text{atoms}}}{2j_{\text{molecules}}}. \tag{4.6}$$

The mechanism explained above is usually called the Eley–Rideal model [4]. The necessary condition for the association of atoms into parent molecules is a good density of atoms adsorbed on the solid surface. The adsorbed atoms are trapped in the potential well on the surface of a solid material. The potential energy of an atom trapped in the potential well is much lower than the potential energy of an atom in the gas phase, as shown in a simplified manner in Fig. 4.7. In reality, the potential is 3D and of a complex structure. Some authors reported binding sites of different binding energies (surface potential wells of different depths); thus the illustration in Fig. 4.7 should be understood just as a very rough picture useful for understanding the basic concepts of heterogeneous surface recombination.

The difference in the potential energies between atoms and the newly formed molecule is transformed into the internal energy of the solid material. The solid material therefore warms up upon exposure to neutral atoms from the gas phase. The molecule formed on the surface according to the Eley–Rideal model may or may not be in thermal equilibrium with the solid material. The potential energy of the atoms in the gas phase is equal to the dissociation energy of the molecule. After desorption, the molecule may keep a substantial part (roughly up to about half of the dissociation energy) of the potential energy in the form of its internal energy [5]. The molecule may, therefore, desorb from the surface in a highly excited vibrational state. Although some research has been performed on the distribution of such molecules over excited states, the phenomenon is still far from being well-understood. A schematic of the Eley–Rideal model is presented in Fig. 4.8.

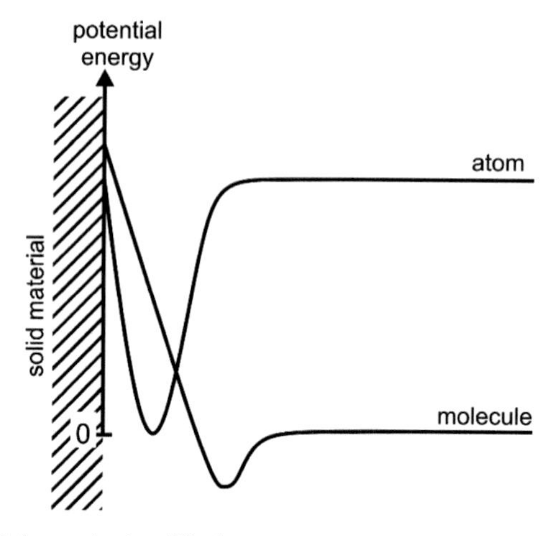

Figure 4.7 Schematic simplified 1D potential curve for atoms and molecules next to the surface of a solid material.

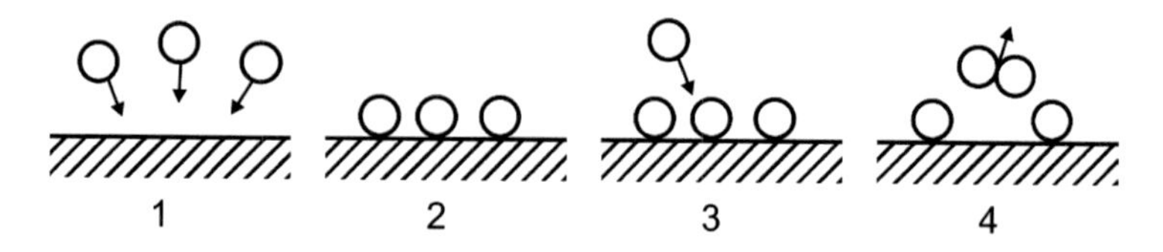

Figure 4.8 Schematic of the Eley–Rideal recombination. (1) Atoms move randomly in the gas phase, (2) atoms adsorb on the surface, (3) an atom from the gas phase interacts with an adsorbed atom, and (4) a molecule desorbs from the surface.

Another possibility is that two adsorbed atoms interact on the surface of the solid material and leave the surface in the form of a molecule. This mechanism is called Langmuir–Hinshelwood and is schematically presented in Fig. 4.9 [4]. In this case, the atoms have already lost their potential energies upon adsorption and, therefore, the molecule cannot leave the surface with a substantial internal energy.

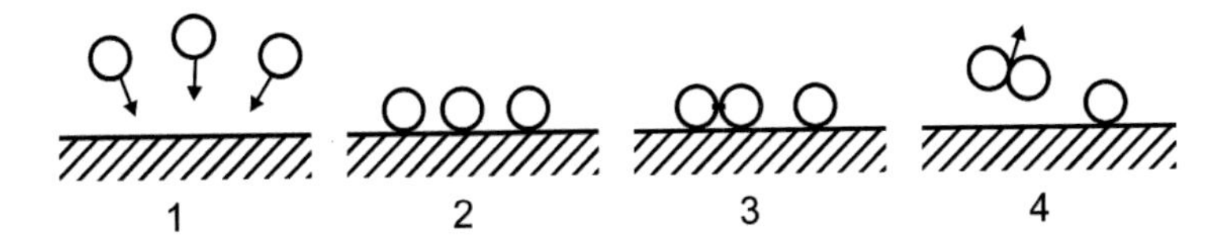

Figure 4.9 Schematic of the Langmuir–Hinshelwood recombination. (1) Atoms move randomly in the gas phase, (2) atoms adsorb on the surface, (3) two adsorbed atoms interact on the surface, and (4) a molecule desorbs from the surface.

The predominant mechanism depends on various factors, from the binding energy to the surface mobility of the atoms and the temperature of the solid material. Experimentally, it is difficult to distinguish between these two mechanisms. In any case, a substantial fraction (usually much more than half) of atoms' potential energy is transformed into the internal energy of a solid material.

The surface recombination causes heating of the solid material exposed to radicals. The heating rate obviously depends on the flux of radicals onto the surface, the dissociation energy of the molecule formed on the surface by association of radicals, and the probability of association, which is expressed in terms of the recombination coefficient, as in Eq. 4.6. For the case of association of atoms to two-atom molecules, the heating power is

$$P = j_{\text{atoms}} \cdot ½W_{\text{d}} \cdot A' \cdot \gamma. \tag{4.7}$$

Here, A' is the surface of the solid material exposed to a flux of atoms j_{atoms}. If a three-atom molecule is formed, the factor ½ should be replaced with ⅓. Combining Eqs. 4.5 and 4.7 gives the heating power

$$P = \frac{1}{8} W_{\text{d}} \cdot n \cdot \langle v \rangle \cdot A' \cdot \gamma. \tag{4.8}$$

Suppose there is a small particle of a simple shape levitating in an atmosphere rich in atoms. The particle is heated at the rate as in Eq. 4.8 and radiates according to the Stefan–Boltzmann law

$$P' = (1-a) \cdot A' \cdot \sigma_{\text{S}} \cdot T^4, \tag{4.9}$$

where a is albedo, σ_{S} is the Stefan–Boltzmann constant, and T is the temperature of the levitating particle. The temperature of the small particle assumes a much higher value than the gas temperature. Taking into account the relations 4.8 and 4.9 and neglecting some effects, the temperature of the levitating particle is

$$T = \sqrt[4]{\frac{W_{\mathrm{D}} \cdot n \cdot \langle v \rangle \cdot \gamma}{8(1-a) \cdot \sigma_{\mathrm{S}}}}. \tag{4.10}$$

The temperature of a levitating particle versus the atom density in its surrounding is shown in Fig. 4.10 for a dissociation energy of 5 eV, albedo 0, and different recombination coefficients γ. In Fig. 4.11 the temperature is shown for the same parameters except that the albedo was taken as 0.9. Here, it should be stressed that several approximations were assumed to simplify the calculations. In real cases, the particle is cooled not only by radiation but also by other effects, including heat transfer because of the thermal conductivity of gas and convection. The approximation is justified only for high differences between gas and particle temperatures. That's why in Figs. 4.10 and 4.11 the curves at lower temperatures are presented with dashed lines.

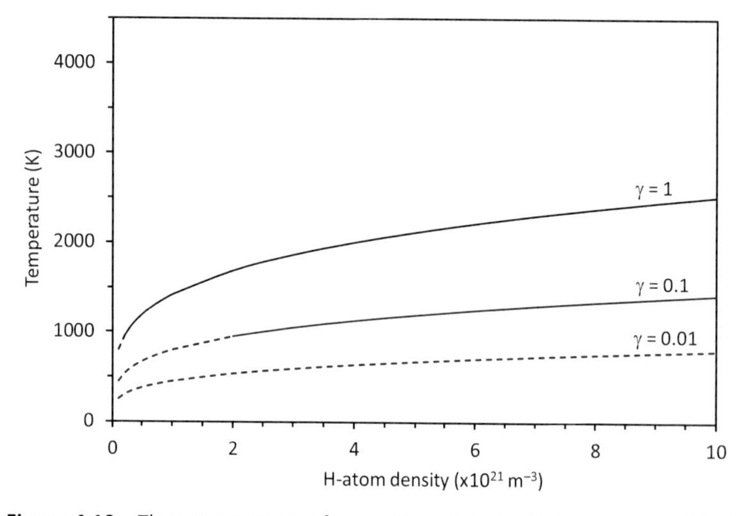

Figure 4.10 The temperature of a small particle levitating in a gas rich in atoms versus the atom density for $a = 0$.

Simple approximations are useful for a rough estimation of the temperature of a small particle. If a particle is large, gradients in the atom density next to the surface occur because of the loss of atoms on the surface. From this point of view, the results summarized in Figs. 4.10 and 4.11 should be used only as the upper values. The approximation is valid for particles up to the dimension of, say, a millimeter. For larger particles, the temperature is lower but still substantially higher than the ambient temperature. The approximation is valid also for a small object immersed in a gas rich in atoms and connected to a thin wire of low thermal conductivity.

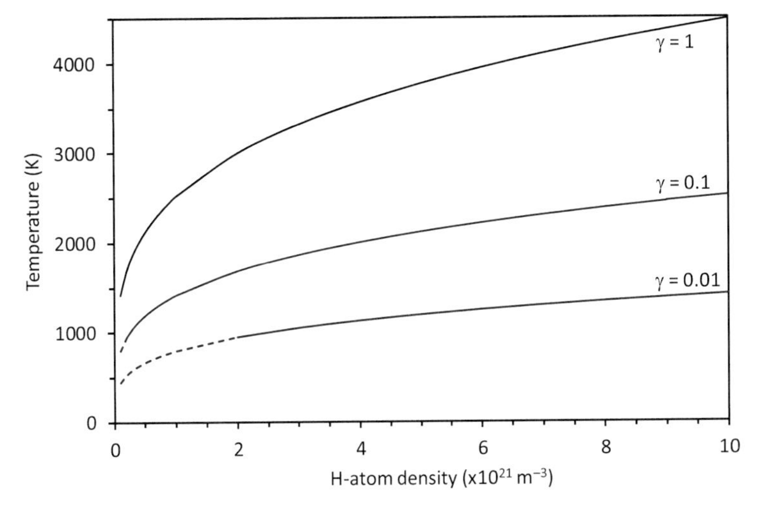

Figure 4.11 The temperature of a small particle levitating in a gas rich in atoms versus the atom density for $\alpha = 0.9$.

Sometimes, it is also useful to estimate the temperature of small particles sitting on the surface of a larger object, provided the thermal connection between such particles and the object is poor, for example, metallic dust on the surface of poor thermal conductors such as glass or polymer. If the small particles sit on the surface, the flux of radicals in Eqs. 4.7–4.10 should be divided by a factor of 2 because the particles are exposed to atoms from one side only.

The heating of a solid material, therefore, depends on the recombination coefficient γ, which has been defined in Eq. 4.6. The recombination coefficient depends on the composition and structure of the surface of the solid material [6, 7]. From this perspective, different groups of materials have been identified.

One group contains materials whose surface has a very limited density of available binding sites for adsorption of gaseous atoms. These materials are often called inert. On the contrary, there are materials of numerous binding sites that are quickly occupied by relatively weakly bonded atoms. Such materials are often called "catalytic" because they serve as catalyzers for heterogeneous surface recombination. Within these two extremes, there are materials of numerous binding sites where atoms bond strongly and materials of moderate density of binding sites capable of binding atoms relatively weakly. The exact picture of available binding sites and shape of potential wells is available only for a few materials of a perfect structure, such as monocrystals, and is still a subject of scientific investigation. In practice, however, the materials do not have perfect properties; therefore, the recombination coefficients have been determined experimentally. For a rough guidance, the materials falling into each category according to available scientific literature and practical experiences of the authors of this text are summarized in Table 4.3.

It should be stressed that the values given in Table 4.3 are just the recommended values for materials of high purity with the native layer of oxide (if applicable) and almost free from other surface impurities. The values are for smooth materials and valid for the range of temperatures not deviating much from room temperature.

While the surfaces of many polymers cleaned by routine techniques are almost free from adsorbed impurities, most metals are quickly covered with a very thin film of impurities in a short time after the cleaning. Many metals form a native layer of oxide whose thickness is often around a nanometer. Furthermore, they tend to adsorb organic impurities, which are always present in air in minute quantities. Both films are rather strongly bonded to the surface of a metallic object and cannot be removed at room temperature even upon prolonged exposure to an ultrahigh vacuum. It is common to detect carbon on metallic samples by surface-sensitive methods under ultrahigh vacuum conditions, such as Auger electron spectroscopy, X-ray photoelectron spectroscopy, and secondary ion or neutral mass spectrometry. The surface impurities often have a strong influence on the recombination coefficient; therefore, a small object placed inside an atom-rich atmosphere does not heat up to elevated temperatures according to Figs. 4.10 and 4.11. The thin film of organic impurities is quickly removed upon exposure to oxygen or nitrogen atoms but may remain on the surface even after prolonged exposure to hydrogen atoms.

Table 4.3 Smooth, clean materials of different recombination coefficients at room temperature for oxygen, nitrogen, and hydrogen atoms

Recombination coefficient	Oxygen	Nitrogen	Hydrogen
Over 0.1	Iron	Iron [8]	Iron
	Nickel [9]	Nickel	Nickel
	Cobalt [10]	Cobalt	Stainless steel
	Copper [10]	Tungsten	Copper
	Silver		Aluminum
			Gold
			Manganese
			Platinum
			Titanium
Between 0.01 and 0.1	Graphite [11]	Aluminum	Graphite
	Stainless steel	Copper	Al_2O_3
	Niobium [10]	Stainless steel	Tungsten
	Tungsten		Cobalt
	Titanium		
	Tungsten		
Below 0.01	Pyrex	Pyrex	Pyrex
	Quartz	Quartz	Quartz
	Teflon [12]	Teflon	Teflon [12]
	Al_2O_3	Molybdenum	PET [12]
	PET [12]		PS [12]
	PS [12]		

The native oxide film does not influence the coefficient for the surface recombination of oxygen atoms because the oxide is formed anyway as a result of chemical interaction of O atoms with metallic surfaces, but it influences the coefficient for hydrogen recombination. H atoms often do not bind much on the surface of an oxidized metal, the bond being definitely much weaker than on pure metals. The values in Table 4.3 are for pure metals. Once a metal with a layer of surface impurities is exposed to H atoms, chemical interaction occurs and the H atoms may eventually etch the layer of organic impurities and reduce the native oxide film. Figure 4.12

illustrates this effect. It shows the dependence of the temperature of small nickel and gold discs versus the fluence of H atoms. For low fluences, the temperatures of both samples remain close to room temperature, which indicates a very small recombination coefficient for these materials covered with surface impurities. After receiving a certain fluence, the temperature starts rising and eventually assumes a steady value typical for catalytic materials. The density of H atoms in the vicinity of the samples was 2×10^{21} m^{-3} in the experiments shown in Fig. 4.12. The behavior as shown in Fig. 4.12 is often called "catalyst activation," and it is explained by the removal of surface impurities. Gold does not form a significant oxide film; therefore, it is activated much faster than nickel. Many metals can be activated in this way, provided the binding energy of oxygen in the film is not too high and the thickness of the oxide film is up to a few nanometers only. Thick oxide films cannot be removed at low temperatures in a reasonable amount of time, and strongly bonded oxides cannot be removed at all by exposure to atomic hydrogen, except at very high temperatures [13–15]. A classic example is aluminum—even the native oxide film will not be reduced on treatment with H atoms unless the temperature is very high. Titanium, on the other hand, serves as a good catalyzer for a heterogeneous surface recombination of hydrogen atoms, provided it is not covered with a compact and thick oxide film [16].

The recombination coefficient also depends on the surface morphology [7]. If the surface is almost perfectly smooth, a substantial fraction of atoms is simply reflected on collision with the solid material. In contrast, rough materials exhibit higher recombination coefficients than their smooth counterparts. The effect is illustrated in Fig. 4.13. In this illustration, an atom is regarded as a rigid sphere. As mentioned earlier, the atoms in this (often not justified) approximation do not feel any attractive force on the surface; therefore, their kinetic energy is preserved and only the normal component of their velocity becomes inverse. Two cases are considered: in the illustration on the left in Fig. 4.13, an atom enters a gap on the surface of a solid material at a large angle. Once it suffers an elastic collision, it is reflected to another side of the gap and experiences several collisions before escaping from the gap. The illustration on the right in Fig. 4.13 shows a rare case where an atom is directed almost perpendicular to the surface. If this happens, the atom suffers only one surface collision. The number of collisions in the rigid sphere approximation obviously depends on the aspect ratio of the gap: if the depth of the gap is much larger than its lateral dimension, the collisions are numerous. The simple drawings are valid for the case when the mean free path is much longer than the gap lateral dimension. Otherwise, an atom also experiences gas phase collisions with other gaseous particles. Collision-less gaps are common in practical cases of rough materials because the mean free path is roughly 1 cm at the pressure of 1 Pa.

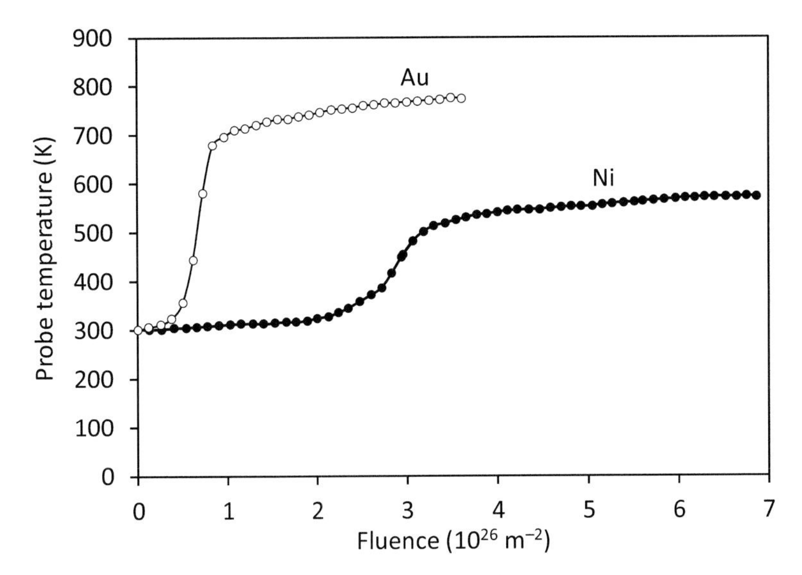

Figure 4.12 Temperature of nickel and gold discs on exposure to atomic hydrogen versus the fluence.

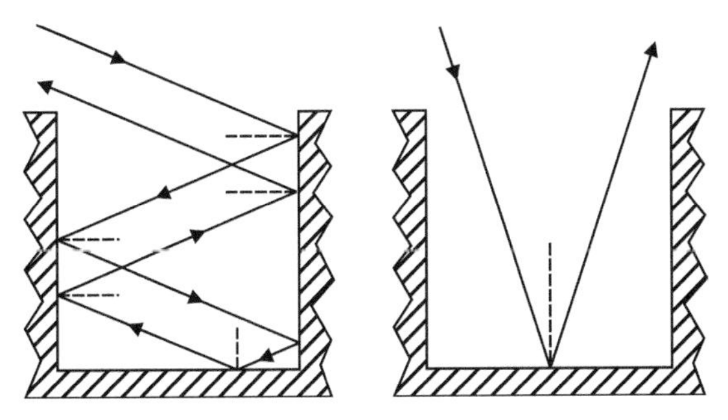

Figure 4.13 Schematic of atom collisions within gaps on a rough surface in the rigid sphere approximation.

The rigid sphere approximation has limited validity. In practice, there is a certain probability that an atom will stick to the surface at each collision. The probability is actually the recombination coefficient for a smooth material as defined in Eq. 4.6. The real recombination coefficient for a rough material is obviously the value for a smooth material multiplied by the number of collisions in the gap. If there are numerous collisions, the recombination coefficient will be large even for materials whose smooth surfaces exhibit a moderate recombination coefficient. In fact, in the approximation of gaps with an infinite aspect ratio separated by infinitely narrow walls, the recombination coefficient for all materials will approach 1. Such materials do not exist, of course, but close approximations are available.

The influence of the surface roughness on the recombination probability of oxygen atoms for the case of graphite is shown in Table 4.4 [11]. The as-received graphite was polished with an abrasive paper with a different grit. One sample of the graphite was made smooth and the other one rough. The surface roughness as measured by a profilometer is shown in Table 4.4 (Fig. 4.14). Smooth graphite exhibits a moderate recombination coefficient, and the measured value for oxygen atoms is 0.04. This means that 96% of the collisions obey the rigid sphere approximation and only 4% result in the adsorption of an atom onto the smooth surface. If the surface is made rough, the atoms suffering an elastic collision are not reflected perpendicularly but most of them experience several collisions before leaving the surface. Because the probability of adsorption at each collision is 0.04 (i.e., the value measured for a smooth surface), the recombination coefficient as determined for a rough sample is much larger, that is, γ = 0.09, for the roughest surface in Table 4.4 [11].

Table 4.4 Dependence of the recombination probability for oxygen atoms on the surface roughness of graphite samples

Sample	Roughness (nm)	Recombination probability
Smooth polished graphite	4.5×10^2	0.04
Rough polished graphite	1.8×10^3	0.06
As-received graphite	3.2×10^3	0.09

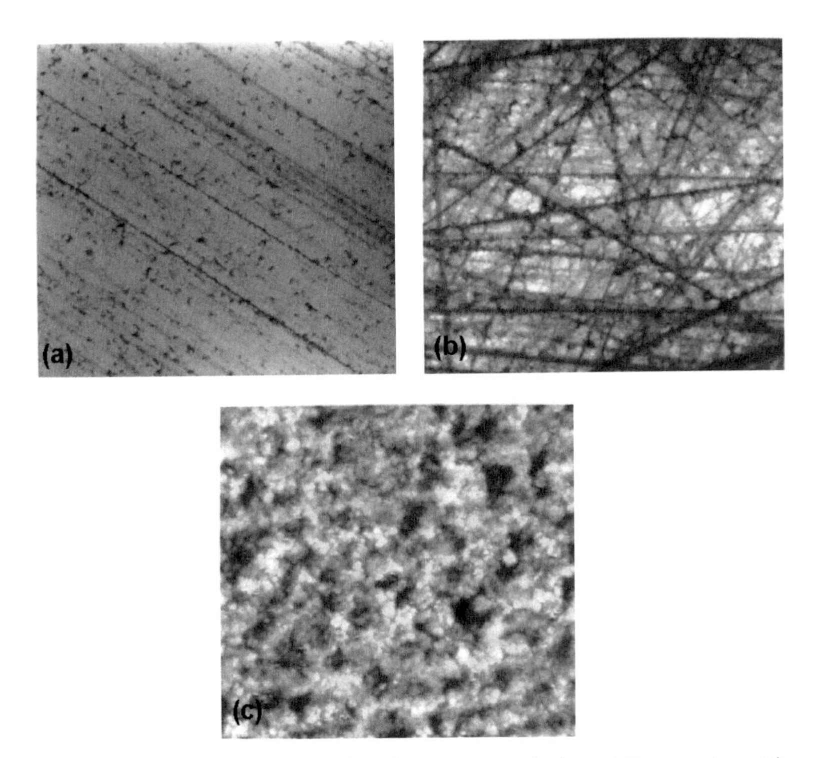

Figure 4.14 Profiles of the surface (1.1 × 1.1 mm) of graphite samples with different surface morphologies and roughness as measured by a profilometer: (a) smooth, (b) rough, and (c) as-received.

Carbon materials of deep and dense gaps separated by thin walls are called carbon nanowalls. The nanowalls are often deposited by the plasma-enhanced chemical vapor deposition process. The organic precursor is dissociated because of the plasma conditions, and the radicals condensate on the surface. In the limited range of plasma parameters, dense nanowalls of vertically oriented multilayer graphene sheets are formed, as shown in Fig. 4.15. Most atoms from the gas phase are trapped in the gaps between nanowalls and suffer numerous collisions before escaping. The effect is illustrated in Fig. 4.15. The measured value of the recombination coefficient for this material is approximately $\gamma = 0.6$ [17].

Even larger recombination coefficients are measured for materials with a very high porosity. Figure 4.16 shows a scanning electron microscope (SEM) image of a nickel foam. The recombination coefficient for oxygen atoms on a smooth nickel foil is 0.27 [9]. In the case of the foam, the atoms are trapped in the open volume within the foam, as shown schematically in the illustration in Fig. 4.16. Very few atoms are reflected on the walls on the surface of the foam because of its high porosity. Most atoms enter the foam and suffer numerous collisions, and very few leave the material. Practically all atoms adsorb on the walls inside the foam, and only molecules are capable of escaping. That is a reason why the recombination coefficient for the nickel foam is practically 1.

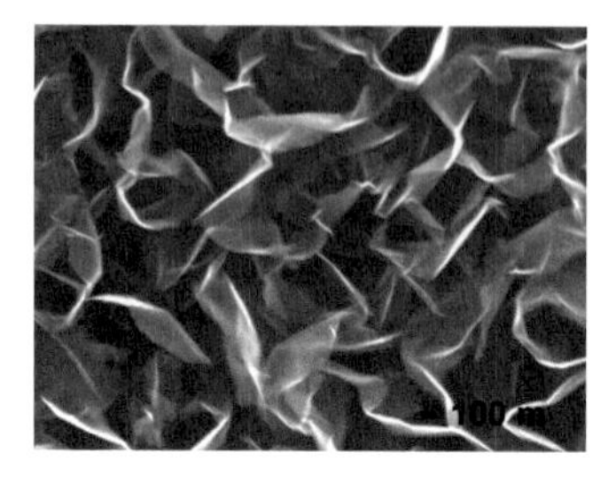

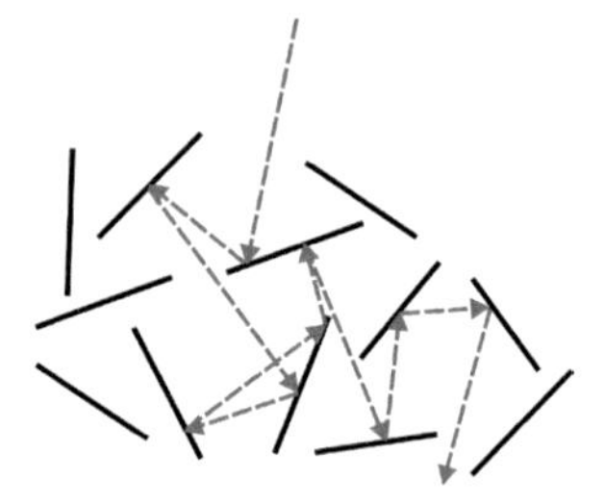

Figure 4.15 SEM image of vertically oriented carbon nanowalls (left) and the illustration of atom collisions (right).

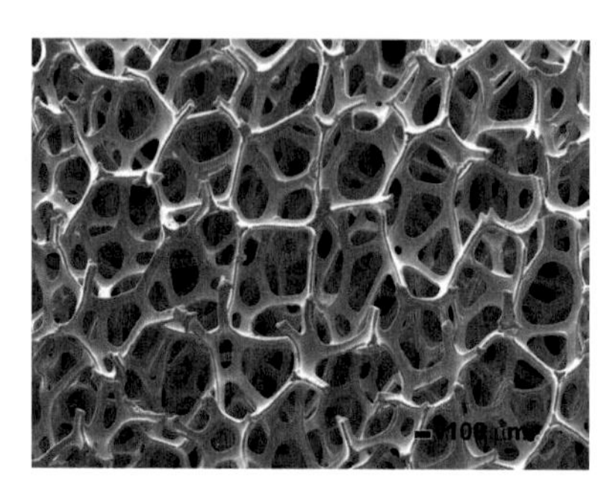

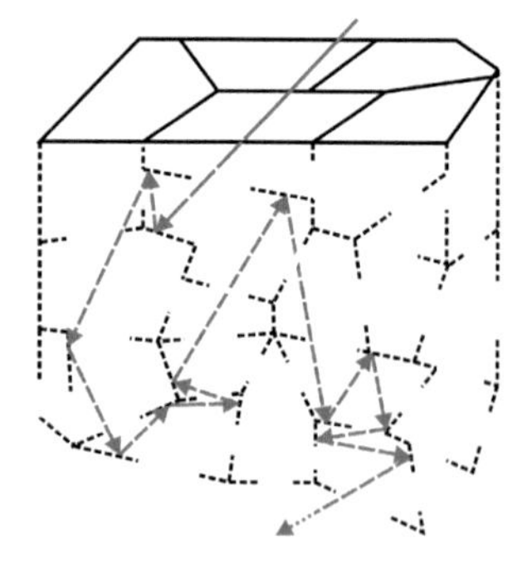

Figure 4.16 SEM image of a nickel foam (left) and the schematic of collisions in the rigid sphere approximation (right).

The recombination coefficients summarized in Table 4.3 were measured at room temperature. Very few experiments have been performed at lower temperatures because of the lack of practical interest. The materials facing a plasma, however, are often heated to high temperatures either deliberately or because of exothermic surface reactions, such as charged-particle neutralization, bombardment with positively charged ions, and, of course, heterogeneous surface recombination of radicals into stable molecules. In some plasmas, the heating of surfaces by absorption of light quanta is not negligible. Any investigation of the loss of atoms on surfaces of plasma-facing materials should, therefore, include thermal effects. Not much work has been done on the temperature dependence of the recombination coefficients for catalytic materials, except for some high-temperature ceramics [18–24]. Figure 4.7 demonstrates that atoms trapped in the potential well cannot leave the surface without associating with a molecule; therefore, the recombination coefficient cannot decrease with increasing temperature. In fact, the recombination coefficient was found to increase with an increasing temperature [18–24]. In some cases, the recombination coefficient increases monotonously with increasing temperature. Figure 4.17 demonstrates this effect. Such monotonous increase could be explained by a higher mobility of atoms adsorbed on the surface. The higher mobility should favor a more intensive recombination following the Langmuir–Hinshelwood model. In some other cases, however, the recombination coefficient was

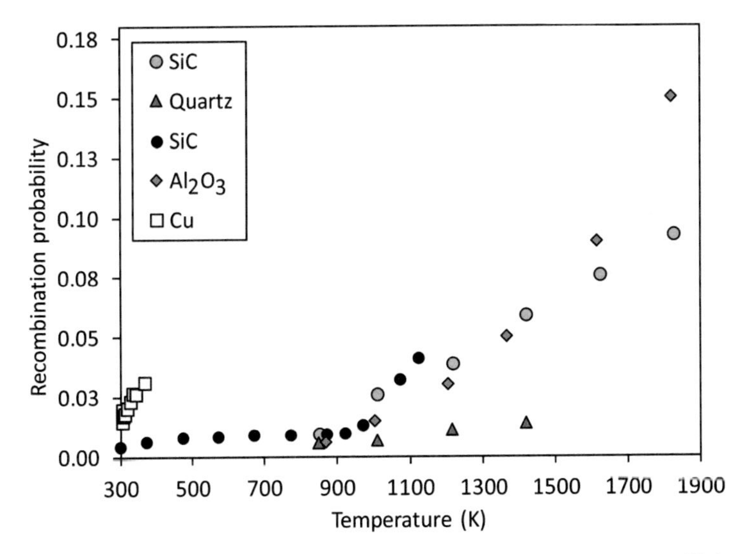

Figure 4.17 The temperature dependence of the recombination coefficient for oxygen atoms on some materials: SiC [18, 20], quartz [18], Al_2O_3 [24], and Cu [23].

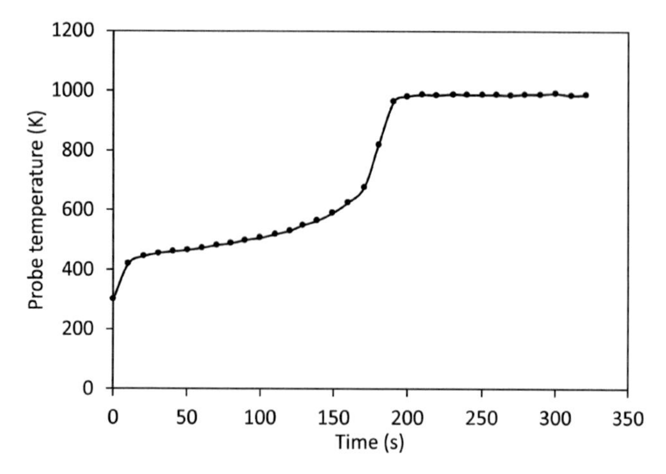

Figure 4.18 The temperature of a small Inconel disc immersed into a H-rich atmosphere of constant density of H atoms of 2×10^{21} m^{-3}.

reported to be almost constant over a broad range of temperatures, indicating the predominant Eley–Rideal model. Interesting results have been found for some alloys. Figure 4.18 shows the temperature of a small Inconel disc immersed in a H-rich atmosphere. The disc was previously activated to avoid the effect shown in Fig. 4.12. One observes three knees in the curve. The first one, at the time of approximately 10 s after the H atom source is turned on, indicates that the material has assumed a constant temperature because the heating and cooling rates have equalized. The temperature, however, does not stabilize completely but keeps increasing slowly for the next 100 s. Such an increase is easily attributed to an increasing gas kinetic temperature but could also be attributed to a slow increase in the recombination coefficient in the range of temperatures between approximately 450 K and 550 K. The second knee is observed at approximately 175 s. Thereafter, the Inconel temperature increases quickly until the third knee is observed, at approximately 1000 K. Such an effect can only be explained by an almost instant increase in the recombination coefficient at approximately 550 K. After the third knee, the temperature stabilizes to an almost perfectly constant value of approximately 1000 K, which is explained by equalization of the heating and cooling rates according to Eq. 4.10. The curve presented in Fig. 4.18 is perfectly reversible. This material, therefore, exhibits two distinct recombination coefficients: a moderately low one at temperatures below approximately 450 K and a high one above approximately 550 K. No scientific explanation for such an unexpected result is available. The effect illustrated in Fig. 4.18 only demonstrates that heterogeneous surface recombination of atoms into stable molecules is still far from being well understood. Therefore, a user of non-equilibrium gases rich in atoms has to take the values presented in Table 4.3 with some caution.

4.2 Plasma Sources of Neutral Atoms

The considerations explained in the previous section are useful and should be taken into account during any attempt to construct an efficient source of radicals, in particular neutral atoms. Although there are several mechanisms that enable the dissociation of molecules into parent atoms, only a couple have been used widely: (i) dissociation on a hot surface and (ii) dissociation because of the plasma conditions. This book refers to the second one. Dissociation occurs in the gas phase at very high gas temperatures, but such a hot source of atoms is not suitable for the treatment of solid materials except materials that remain stable at high temperatures. Technologically more important is the dissociation of gases such as gaseous plasma at a low gas temperature in non-equilibrium states sustained at a low power density. Here, it should be stressed that the term "gas temperature" is not defined in highly non-equilibrium gases. Many authors have adopted the expression "neutral gas kinetic temperature" or just "gas kinetic temperature," which indicates the temperature arising from the random motion of gaseous particles. The expressions "electron temperature" and "ion temperature" indicate the temperature arising from the random motion of electrons and ions, respectively. A few authors have adopted the expressions "dissociation temperature" and "ionization temperature," which actually indicate the dissociation and ionization fractions, respectively. Most authors rather indicate the densities of neutral radicals and charged particles.

Following the considerations explained in Section 4.1 enables the construction of efficient radical sources. Electron impact is the most straightforward method for the dissociation of molecules. A gaseous plasma is rich in electrons, and their temperature is much higher than the neutral gas kinetic temperature. If an electron is energetic enough (i.e., its energy is higher than the dissociation threshold), an inelastic collision with a molecule may cause dissociation. The probability of such an event is really small at the threshold and increases with increasing electron energy, peaking at approximately 100 eV. Such high-energy electrons are really scarce in common plasmas. Therefore, the probability of direct dissociation of a molecule from the ground state is low. The dissociation energies of many molecules are rather large at values between approximately 5 and 10 eV, but many molecules have metastable states that help increase the dissociation efficiency. For example, oxygen molecules have two metastable states at the excitation energies of roughly 1 and 2 eV. The dissociation energy of oxygen molecule is 5.16 eV. Thus, intermediate excitation to those metastable states effectively lowers the required electron energy, what is useful, especially because the electron energy distribution function is usually almost exponential in the high-energy tail.

The dissociation fraction of molecules in a gaseous plasma definitely increases with increasing discharge power, but so does the energy consumption; therefore, increasing the power is not the best way for ensuring a high dissociation fraction. More important is minimizing the loss rate, which is either in the gas phase due to three-body collisions following Eq. 4.2 or on surfaces through heterogeneous recombination. The efficiency of the surface loss of atoms depends on the recombination coefficient—see Table 4.3. There should be a maximum density of atoms in a gaseous plasma at an optimal pressure. At pressures below the optimum, the atom density should decrease because of surface effects, whereas at pressures above it, the atom density should decrease because of the loss of atoms in the gas phase. Figure 4.19 illustrates the effects. The pressure at which the optimum occurs depends on the discharge power density: at larger powers the optimum is shifted to larger pressures.

Figure 4.19 clearly demonstrates that high-pressure discharges are inefficient sources of atoms. In fact, the specific power needed to sustain a large density of atoms at atmospheric pressure is huge compared to the low-pressure counterparts. For example, a simple atmospheric-pressure discharge operates at a volume on the order of 1 cm^3 at a power on the order of 10 W, giving a power density of 10 W cm^{-3} = 10^7 W m^{-3} and an atom density on the order of 10^{21} m^{-3}. The same atom density in low-pressure chambers is obtained at power densities of 10^3–10^4 W m^{-3}. The power used for sustaining a gaseous plasma is used for different reactions, but practically, all power is finally transformed into heat because a system that is in thermodynamic non-equilibrium approaches equilibrium conditions spontaneously soon after the discharge power has been turned off. The atmospheric-pressure atom sources are thus energetically inefficient.

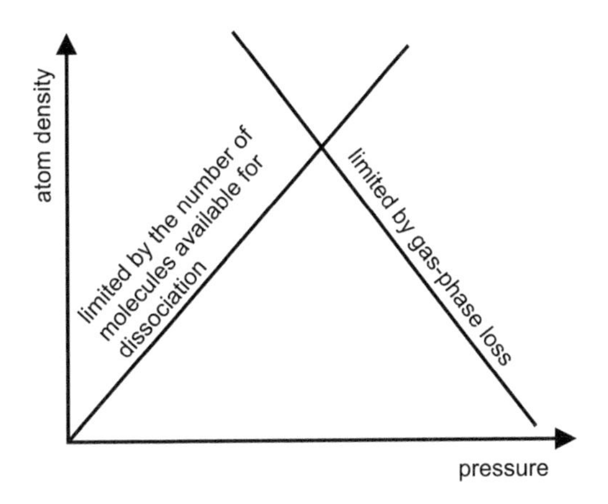

Figure 4.19 Schematic of the O atom density versus pressure.

Low-pressure sources of radicals employ different gaseous discharges. High-power discharges are often avoided because they cause increased neutral gas kinetic temperature and are therefore useful only for the treatment of materials that can stand high thermal loads. Low-power discharges are divided into direct current (DC) and low-frequency alternate current, and high-frequency discharges are divided into radio frequency (RF) and microwave (MW) discharges. DC and low-frequency discharges are rarely used as radical sources because they employ two electrodes inserted into the plasma chamber. Electrodes are made from electrically conductive materials such as metals and graphite. These materials exhibit rather high catalytic activity for surface recombination of atoms (see Table 4.3). Therefore, they are often avoided when constructing radical sources.

High-frequency sources are common. Gaseous plasma is sustained by low-frequency RF discharges (frequency on the order of 10 kHz), high-frequency RF (roughly from 1 to 100 MHz), or MW discharges (on the order of GHz). Particularly popular are industrial frequencies of 13.56 MHz for RF discharges and 2.45 GHz for MW discharges and their harmonics.

A low-frequency RF discharge usually employs an electrode immersed into the grounded housing and powered with a generator. A schematic of such a setup is shown in Fig. 4.20. The metallic chamber as well as the electrode definitely represent a sink for radicals because of a heterogeneous surface recombination. One solution of this problem is a dielectric coating of the metallic components facing the plasma. The coating is made from a material preferably of a low recombination coefficient, for example, a thin layer of quartz glass. The layer should be thin enough to ensure a reasonable impedance. The impedances involved in the configuration presented in Fig. 4.20 are shown schematically and in a simplified manner in Fig. 4.21. The total impedance is the sum of capacitive and resistive components. The capacitance occurs across both the glass film and the sheath between the glass surface and the plasma. The capacitive component of impedance is inversely proportional to the frequency and capacity of the plane-parallel capacitor:

$$Z_{\mathrm{C}} = \frac{1}{\omega \cdot C} = \frac{1}{\omega \cdot \left(\varepsilon \cdot \varepsilon_0 \cdot A''/x\right)}. \tag{4.11}$$

Here, ω is the frequency of the power supply, ε is the relative permittivity, ε_0 is the vacuum permittivity, A'' is the area of the capacitor, and x is the thickness of either the thin dielectric film or the sheath. The capacitive component of the impedance is thus inversely proportional to the electrode area. To minimize it, one should apply an electrode as large as possible. Obviously, if one wants almost all the discharge power to be dissipated on the powered electrode, the electrode should be small. In the case of radical sources, small electrodes should be avoided to minimize the power lost on heating of the electrode. The capacitive component of the impedance is proportional to the thickness of the dielectric layer; therefore, the glass film should be made as thin as possible.

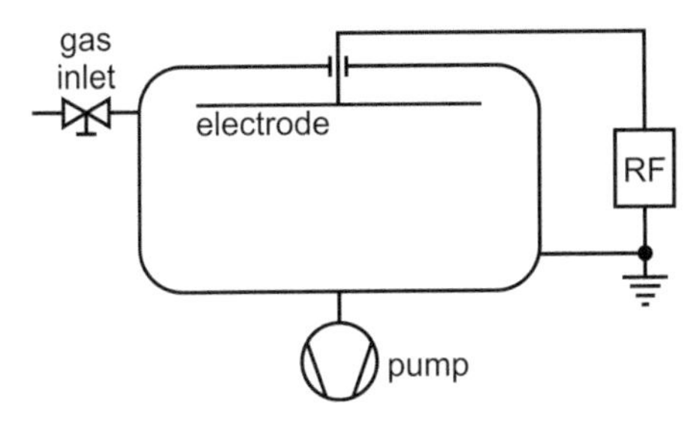

Figure 4.20 Schematic of the capacitive-coupled RF discharge with an electrode immersed into the plasma reactor.

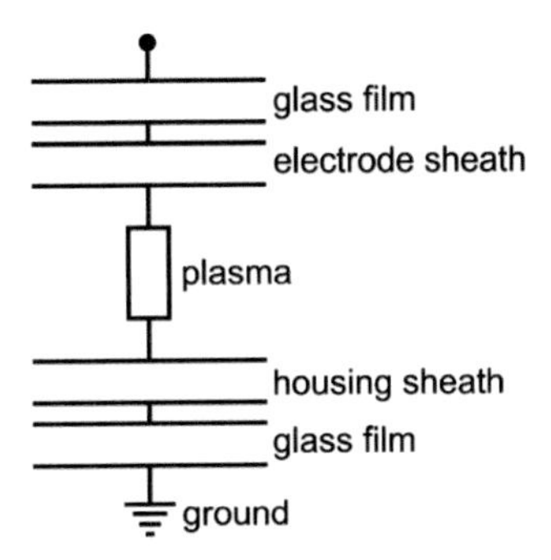

Figure 4.21 Simplified electric circuit for the configuration presented in Fig. 4.20.

There is substantial impedance across the plasma sheath next to the electrode. The sheath thickness depends on the Debye length, which in turn depends on the electron density and temperature as follows:

$$\lambda_D = \sqrt{\frac{\varepsilon_0 k_B T_e}{e_0^2 n_e}}, \tag{4.12}$$

where ε_0 is vacuum permittivity ($\varepsilon_0 = 8.85 \times 10^{-12}$ F/m), e_0 is the elementary charge, T_e is the electron temperature, and n_e is the electron density.

The fact that the capacitive component of the sheath cannot be neglected is because of the poor electrical conductivity of the sheath as compared to that of the plasma. In a gaseous plasma, the current is carried by electrons that have excellent mobility because of their low mass. In the sheath, the density of the electrons is much lower than in plasma because of a self-biasing effect (the electrode is highly negative against plasma). Therefore, the current carriers in the sheath are positively charged ions of limited mobility.

The same effects as for the powered electrode also occur on the grounded housing, but the surface (A″) is usually larger. Therefore, the impedance of capacitors at the bottom of Fig. 4.21 is smaller than that of the top capacitors. The capacitive component of the impedance is easily suppressed when the frequency of the RF generator is increased, as demonstrated by Eq. 4.11. Still, low-frequency RF discharges are popular as sources of specific radicals because of the low cost of low-frequency RF generators and because of a highly effective transfer of the generator's power to the gaseous plasma—no complex matching network is needed.

Low-frequency RF discharges are particularly useful as sources of radicals containing several atoms. They are widely used as sources of radicalized monomers. A monomer is introduced through a gas inlet, as shown schematically in Fig. 4.20. It radicalizes because of the plasma conditions, and the radicals stick to the surface of any material immersed into the plasma reactor. Low specific power (often around 1 kW m^{-3}) is applied to enable only partial radicalization of the monomer, which in turn enables the preservation of the monomer's original composition and structure. The process is often called "plasma polymerization" and is used nowadays at the industrial level.

Such low-power discharges are useful for the partial radicalization of molecules containing numerous atoms but are impractical for the efficient dissociation of molecules into the smallest fragments—atoms. The high dissociation fraction is obtained at larger power densities and frequencies. Particularly popular sources of neutral atoms are electrodeless high-frequency RF discharges. The schematic of a simple setup widely used on an experimental level is presented in Fig. 4.22. A dielectric tube is made from a material of a low recombination coefficient. Considering Table 4.3, the choice of materials includes some glasses, ceramics, and polymers. Because the polymers often interact chemically with plasma radicals, they are usually avoided. The most frequently used materials for the discharge tube are borosilicate and quartz glasses. Some types of borosilicate glass can be sealed with some metals (the allow Kovar is particularly useful) to make perfectly hermetically tight joints free from gaskets. An alternative to sealing of the flanges to the glass tube is the application of gaskets made from elastomers. Borosilicate glass is practical because it can easily be blown into an arbitrary shape and exhibits a low recombination coefficient for many atoms, including oxygen and nitrogen. For hydrogen, however, it should be avoided because the recombination coefficient is moderate and increases with increasing glass temperature. One advantage of borosilicate glass is effective absorption of UV radiation, especially hard UV radiation. If radiation penetrates the glass, it represents a health risk. Not only is it harmful for the eyes, but hard UV radiation also causes the formation of ozone in the laboratory. From this point of view, borosilicate glass exhibits a definite advantage over some other types of glass, such as quartz.

Quartz glass is also popular for discharge chambers in the configuration presented in Fig. 4.22. It is a particularly useful material for the construction of sources of hydrogen atoms because the recombination coefficient is rather low. A user of quartz, however, should be wary of its transparency to UV radiation. Hydrogen plasma is a source of continuum radiation in the range of UV wavelengths arising from the transition of the triplet H_2 molecules from highly excited states to the dissociative state. The photon energy is high enough to dissociate oxygen molecules, which results in the appearance of ozone in the laboratory.

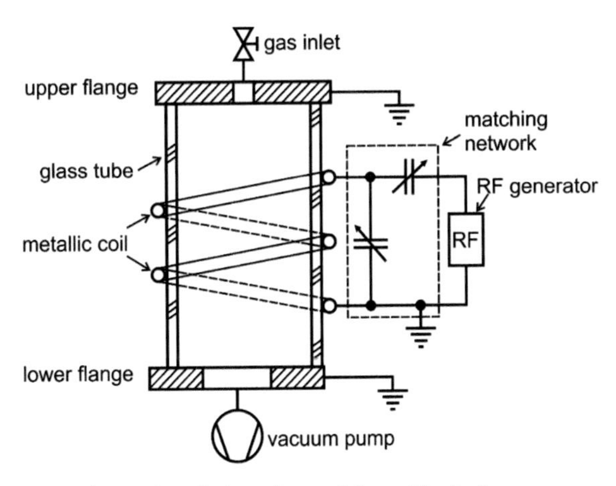

Figure 4.22 Schematic of the electrodeless RF discharge powered by an inductor.

RF power couples with the plasma in Fig. 4.22 through a coil. The coil is connected to an RF generator operating at high frequencies, often 13.56 or 27.12 MHz, via a matching network. The network may consist of variable capacitors and/or coils, but typically a couple of capacitors are used, as shown in Fig. 4.22. One capacitor is connected in series and the other in parallel. The matching network is essential because the plasma impedance depends on the plasma density; therefore, the electrical circuit should be matched for each discharge condition. If not, the generator would be mismatched and the majority of the available power would be just reflected.

At the low discharge power used in the configuration presented schematically in Fig. 4.22, the basic concept of the electrical circuit applies, but with an important modification that includes the influence of both flanges. One such modified circuit is presented in Fig. 4.23 in a simplified manner.

The circuit presented schematically in a simplified manner in Fig. 4.23 shows an important difference in the capacitances next to the coil and flanges. Next to the coil, two capacitances occur in series: one is attributed to the capacitance of the sheath between the bulk plasma and the inner surface of the glass tube, and the other capacitance is attributed to the capacitance of the glass wall. The capacitive component of the impedance is inversely proportional to the capacitance of a capacitor as in Eq. 4.11. The thickness of the glass tube is usually larger than the sheath thickness because of practical reasons—the glass tube should withstand the force as a result of the pressure difference, which is approximately 1 bar. Next to the flanges, there is only the capacitance owing to the sheath, because the plasma is in direct contact with the metallic flange. The absence of impedances because of the absence of the glass wall next to the flanges has an important consequence: The impedance of the circuit from the powered part of the coil and the grounded flanges is lower than the impedance of the circuit from the powered part of the coil to the grounded part of the coil despite the rather large distance between the coil and the flange. As a result, a uniform plasma expands throughout the glass tube. It should be stressed again that the considerations presented in this paragraph are valid only for the case of low power where the density of the charged particles in the gaseous plasma is low. In scientific literature, such a configuration is often called "inductively coupled plasma in E-mode." "E" indicates that the electrons gain energy in the electric field next to the sheath rather than in the induced electric field that appears because of an oscillating magnetic field inside the coil.

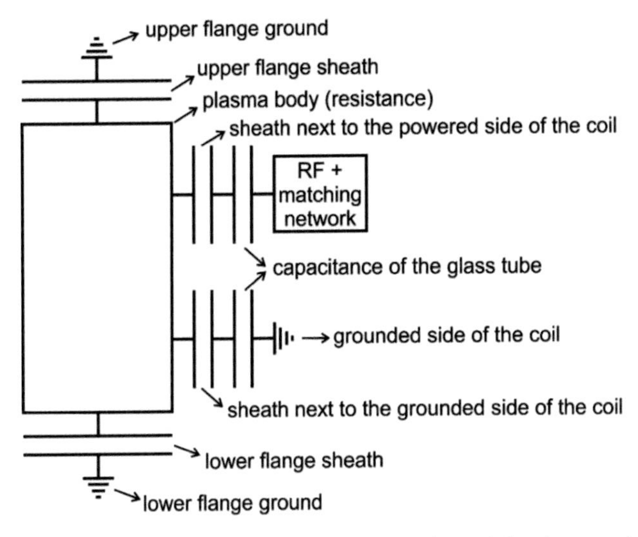

Figure 4.23 Schematic of the simplified electrical circuit for the experimental setup presented in Fig. 4.22 when the discharge power is low (plasma in the E-mode, capacitive coupling).

In the case of high available power, the circuit presented in Fig. 4.23 is not valid. At high power, the electron density is high enough for the plasma to be presented as a highly conductive medium. In such a case, the circuit shown in Fig. 4.23 should be replaced with a completely different one that includes the inductance of the gaseous plasma, which becomes important and eventually predominant when the electron density is large. A suitable circuit for such a case is presented in Fig. 4.24 in a simplified manner. The coil connected to the RF generator via the matching network (see Fig. 4.22) represents the primary coil of the transformer. The secondary coil is the plasma, which is coupled with the primary coil by the mutual inductance that occurs when the current in one inductor induces a voltage in another, nearby inductor. The gaseous plasma is, of course, inside the primary coil. Such a picture is justified for charged particles of a high density. The plasma also has resistivity (ohmic impedance), which is represented by a resistor connected in parallel with the plasma's inductance. A highly conductive plasma following the scheme in Fig. 4.24 is often called "inductively coupled plasma in H-mode." "H" indicates that the electrons gain energy in the induced electric field inside the coil, which is a result of an oscillating magnetic field.

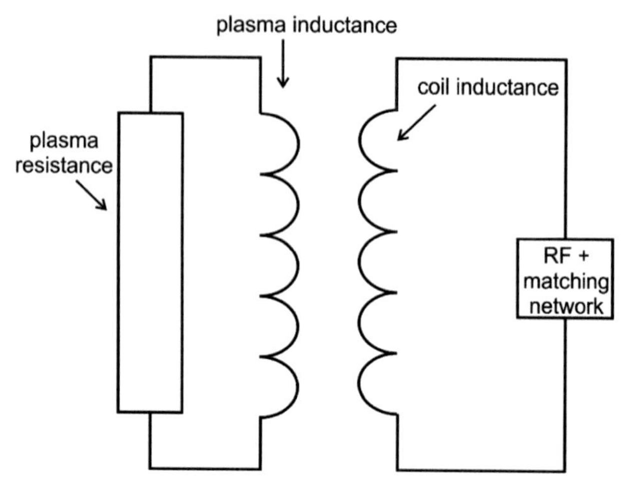

Figure 4.24 Schematic of the simplified electrical circuit for the experimental setup presented in Fig. 4.22 when the discharge power is high (plasma in the H-mode, inductive coupling).

The simplified picture presented in Fig. 4.24 neglects the capacitive components of the circuit impedance because of the presence of the glass tube, which is usually not justified. Therefore, it should be regarded only as a rough approximation. Still, it is useful to understand the huge difference in the mechanisms of the inductive RF plasma powering in E- and H-modes. It is also worth mentioning than the impedance, which is a result of the capacitance of the sheath between plasma and a surface, is neglected because the electron density is large and therefore the Debye length (Eq. 4.12) is small and so is the impedance.

A plasma can be sustained in the H-mode only when the power density is large enough to ensure a high electron density. The high power is attributed to high voltages on the coil. Such high voltages cause intensive radiation (the coil acts as an antenna), which represents a health hazard. Furthermore, the high voltages may damage the capacitor in the matching network and also cause increased capacitive coupling to nearby grounded metallic components. For large plasma reactors, the voltages may be too high and therefore alternative couplings are recommended. One possibility is the application of several coils connected in parallel to decrease the voltage and enable expansion of the coil length. This solution is schematically presented in Fig. 4.25. In this case, the distance between the coils should be large enough to prevent predominant electric current between neighboring coils. Otherwise, the configuration in Fig. 4.25 can be represented as a large cylinder placed onto the glass tube. In practice, the coils should be a few centimeters apart.

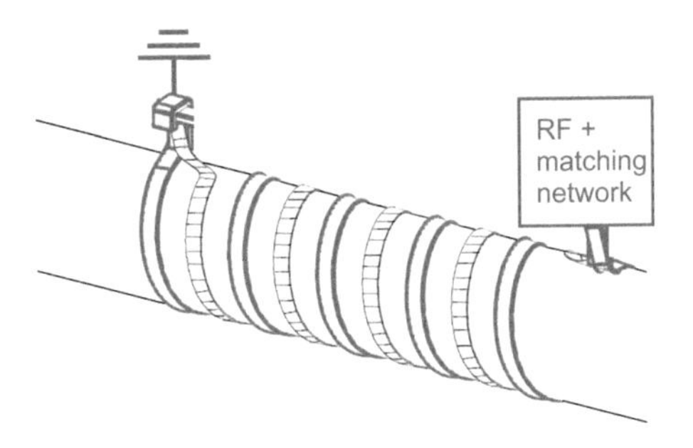

Figure 4.25 Schematic of two coils connected in parallel, suitable for an industrial-size inductively coupled RF plasma in the H-mode.

Yet another coupling suitable for sustaining plasmas in long cylindrical discharge tubes is presented in Fig. 4.26. This configuration benefits from quadrupole coupling. There are four rods placed symmetrically along the glass tube. Two opposite rods are powered, and the other two are grounded. The coupling can be either in the E-mode (predominant capacitive coupling, as schematically shown in Fig. 4.21) or in the H-mode (predominant inductive coupling, as schematically shown in Fig. 4.24). Obviously, the capacitive coupling will be established at low power of the RF generator and inductive at high power. Another important parameter is the diameter of the discharge tube. The coupling presented in Fig. 4.26 is effective in the H-mode at a rather small diameter, say up to approximately 10 or 20 cm. In the configuration presented in Fig. 4.26, the rods connected to the RF generator represent almost perfect antennas. Therefore, the electromagnetic radiation is never negligible. If such a configuration is adopted, the plasma reactor should be caged in a metallic shield.

Obviously, the distance between the rods and the shield should be large (over 10 cm) to minimize the capacitive ground of the powered rods to the metallic shield.

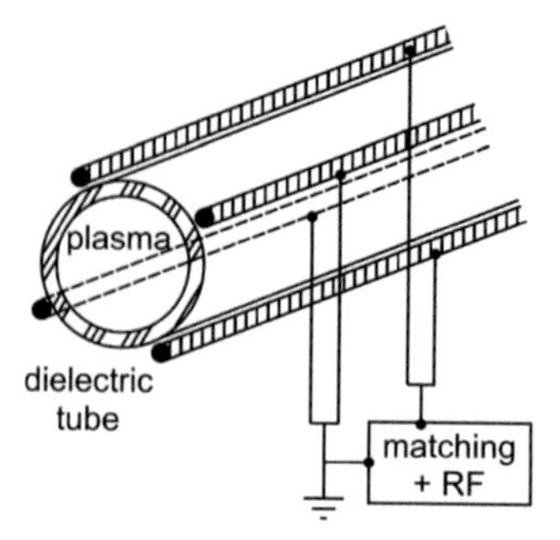

Figure 4.26 Schematic of coupling using a quadrupole coupling.

The configurations presented in Figs. 4.25 and 4.26 are useful for powering long and rather narrow plasma reactors. In many practical examples, however, reactors of a large diameter are preferred. In such cases, a different coupling mechanism between the RF generator and the gaseous plasma should be applied. Three possible configurations are shown in Figs. 4.27–4.29.

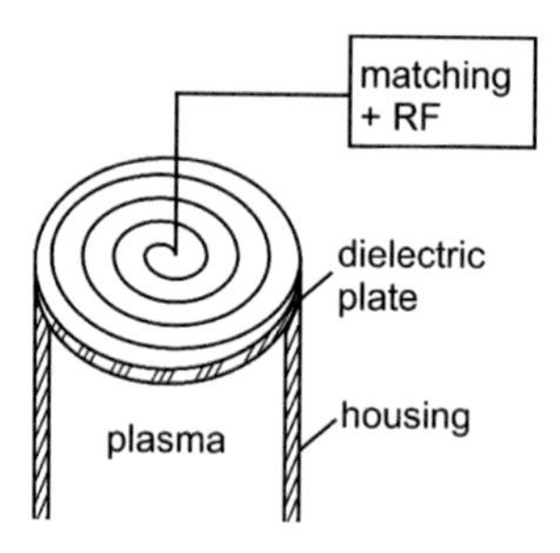

Figure 4.27 Schematic of coupling using a spiral.

Figure 4.27 shows coupling suitable for the sustenance of gaseous plasma in a low-pressure reactor using a spiral. The spiral is placed onto a dielectric plate and connected to the RF generator via a matching network at the center of the system, as shown in Fig. 4.27. The other side of the spiral may or may not be grounded with an electrical conductor. If it is not grounded deliberately, the electrical circuit is closed via capacitive coupling with the metallic chamber or through the gaseous plasma, or the spiral just represents an antenna. Such a configuration is useful for powering plasmas in reactors of large diameters, but one should consider the limitations arising from the simple fact that a long conductor also represents a high inductive impedance. Too long spirals will represent such a high impedance that the voltage across the spiral would be detrimental to the capacitor inside the matching network. Furthermore, the dielectrics are often fragile, so if the diameter of the dielectric plate is too large, the pressure difference would cause an extremely large force and the dielectric would break. Here, it is worth mentioning that the pressure difference of 1 bar is 10 N/cm^2. If the area of the dielectric plate is 1 m^2, the force because of the pressure difference would be equivalent to a mass of 10 tons. The configuration presented in Fig. 4.27 is, therefore, suitable for powering plasmas in medium size reactors only. If the dielectric plate dimension exceeds approximately 1 m^2, it should be straightened mechanically and the spiral should be replaced with several parallel coupled spirals using the same philosophy as presented schematically in Fig. 4.25.

Another solution to a reasonably uniform plasma in a wide, rather short chamber is presented in Fig. 4.28. In this case, the upper plate is metallic, so one benefits from the toughness of metals. The plate can be further reinforced by ribs (not shown in Fig. 4.28). The plasma is sustained in several inductively coupled discharges connected in parallel and powered with one powerful RF generator. Each discharge operates in a glass tube. For the sake of clarity, only two such discharges are shown in Fig. 4.28, but in practice, several individual discharges may be used. The impedances of particular discharges vary; therefore, it is good to distribute the available power evenly to all discharge tubes. An important characteristic of gaseous plasma is that its electrical conductance increases with increasing plasma density; therefore, if a particular discharge creates a plasma of a higher electron density than others, its impedance will decrease, causing more current to flow through the denser plasma, which further decreases its impedance, and so on. Eventually, the current will flow through one discharge only. Fortunately, the impedance does not decrease with increasing plasma density infinitely because other effects suppress it. The most important one is the skin effect: if the plasma becomes very dense, the electromagnetic field does not penetrate deep into the conductive medium. Therefore, the impedance increases with further increase of the plasma density. As a result, the configuration shown in Fig. 4.28 remains suitable for powering plasma in the entire reactor. At a low power of the RF generator, the plasma is ignited in one discharge tube. As the power is increased, the plasma also appears in another discharge tube, and eventually, at a large RF power, the plasma is sustained in all discharge tubes. Such an effect is actually achievable, provided the impedances of the connecting lines are adjusted properly. In practice, it is difficult to match all impedances involved in the electric circuit containing numerous discharges; therefore, the uniform density of the gaseous plasma in the reactor presented in Fig. 4.28 remains a technological challenge.

A method to suppress the above-mentioned effect is presented schematically in Fig. 4.29. Here, the individual inductively coupled discharges are connected with a dense plasma, which, as mentioned, has a rather low impedance at a high electron density. Although the technological problem of equilibration of the total impedances of separate discharges still persists, it is less severe than in the configuration presented in Fig. 4.28. The configuration in Fig. 4.29 usually contains numerous separate inductively coupled discharges, creating a reasonably uniform plasma in the entire reactor. A drawback of the configuration in Fig. 4.29 is the appearance of a dense gaseous plasma in the entire glass tube outside the main chamber, because of which the energy efficiency of such a configuration is rather inadequate.

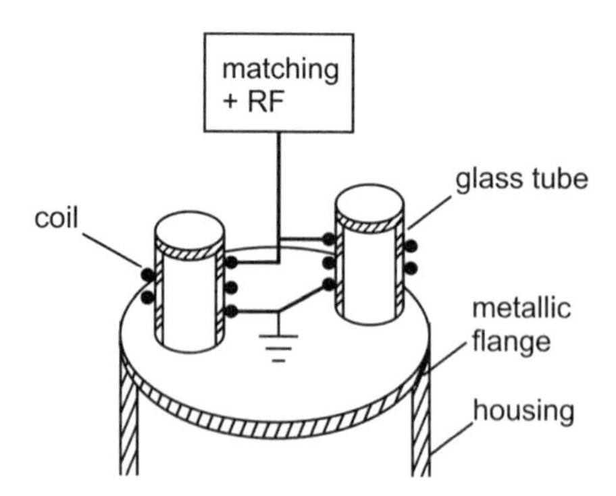

Figure 4.28 Schematic of coupling using multiple (two) inductively coupled discharges.

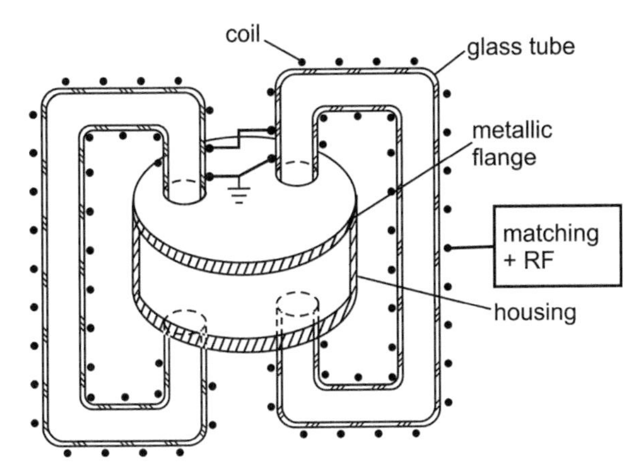

Figure 4.29 Schematic of coupling using multiple inductively coupled discharges suitable for narrow reactors.

All configurations of the inductive RF coupling as described above employ an antenna or inductor placed outside the vacuum chamber. The reason is that if a coil is placed inside the reactor, the electrical circuit would be closed through the element of least impedance, that is, the dense gaseous plasma. As a result, the powered side of the coil would be subjected to an intensive heat load and arcing would occur. This effect, however, can be suppressed rather effectively using very low-impedance antennas immersed into the plasma reactor. Figure 4.30 represents the schematic of a plasma reactor employing numerous low-impedance antennas. Each antenna is actually a U-shaped conductor fully covered with a thin glass or ceramic film for dielectric isolation from the plasma. The typical impedance of a single antenna is on the order of 10 Ω only. The antennas are connected in parallel to the RF generator, as shown schematically in Fig. 4.30. Such a configuration allows for efficient high-density plasma production and low potentials between the powered sides of antennas and the grounded housing. The problems of adjusting impedances of connecting elements still persists, but careful matching allows for a rather uniform plasma in the entire volume. The configuration in Fig. 4.30 is scalable to very large plasma reactors, provided the generator is capable of delivering very large electrical currents.

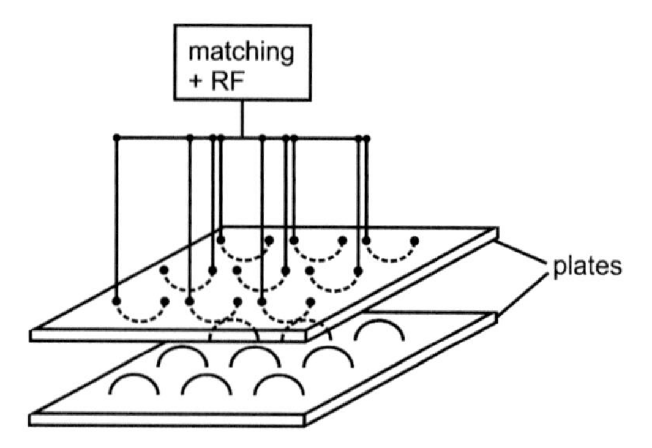

Figure 4.30 Schematic of a plasma reactor with numerous low-impedance antennas immersed inside a vacuum chamber.

Inductively coupled plasma powered with high-frequency RF generators is probably the most common source of neutral atoms. Another option is the application of MW discharges, which are particularly suitable for the sustenance of a gaseous plasma in a small volume. When the frequency of the electromagnetic field is in the range of gigahertz, the skin effect cannot be neglected. The penetration depth of the electromagnetic field in a conductive medium decreases as the square root of the electrical conductivity. Therefore, the heating of plasma electrons is limited to a thin plasma layer next to the surface. The most popular atom sources are plasmas created in glass tubes approximately 1 cm in diameter. The dissociation efficiency is large, but the plasma is never really cold in such discharges.

4.3 Production of O, H, and N Atoms in an Inductively Coupled RF Discharge

This section gives a practical example of the production of neutral atoms in a gaseous plasma sustained by an inductively coupled RF discharge. As explained in the previous sections, the electrodeless configuration of the high-frequency RF discharge is particularly suitable for the efficient dissociation of oxygen, nitrogen, and hydrogen molecules into parent atoms. The reason for the high efficacy is twofold: (i) the RF power is efficiently absorbed by the electrons in the gaseous plasma, and (ii) the loss of atoms by heterogeneous surface recombination is minimal.

In the example elaborated here, the RF power is transferred to the gaseous plasma either by the predominant capacitive coupling (schematic of the equivalent electrical circuit is shown in Fig. 4.23) or by almost pure inductive coupling (as shown schematically in Fig. 4.24). The RF generator operates at the most frequently used industrial frequency of 13.56 MHz up to the nominal power of 1000 W. There are two power meters on the RF generator: one measuring the forward power and the other indicating the reflected power. If the reflected power is larger than 200 W, the generator shuts down automatically to prevent any damage. The predominant capacitive coupling will be called "E-mode," whereas the expression "H-mode" will be assigned for the predominant inductive coupling. The matching network in the example shown in this section was optimized for the coupling in the H-mode, enabling high discharge powers and thus a high density of electrons.

The experimental system will be described in detail, enabling a reader to replicate the results. The schematic of the experimental system is shown in Fig. 4.31. The discharge chamber is a tube that is 80 cm long and has outer and inner diameters of 40 and 36 mm, respectively. The tube is made from borosilicate glass of a recombination coefficient of approximately 10^{-3} for either O or N atoms, whereas the recombination coefficient for H atoms is closer to 10^{-2} at room temperature. The coefficient for H atoms increases with increasing glass temperature [25]. The glass tube is connected to standard KF flanges using elastomer gaskets. The RF coil is mounted asymmetrically onto the discharge tube. This was done intentionally to facilitate different couplings to the flanges. The coil of 6 turns is made from a copper tube of outer and inner diameters of 3 and 1.5 mm, respectively, and is connected to an RF generator Cesar supplied by Advanced Energy. The coil is cooled with water, whereas the segment of the discharge tube with the intensive plasma is forced-air cooled. Such cooling prevents heating of the outer walls of the discharge tube over approximately 50°C even for large discharge powers. Gases are introduced into the discharge tube through an Aera FC-7700 mass flow controller. An absolute vacuum gauge, the MKS Baratron 722A Absolute Pressure Transducer is used for measuring the pressure inside the discharge tube, which is pumped by an Edwards two-stage rotary pump of a nominal pumping speed of 80 m^3/h. A valve IEVT KV-R-32 supplied by Vacutech enables the separation of the pump from the discharge tube without turning it off. There is also a flow restrictor above the vacuum pump, which enables the adjustment of the effective pumping speed. Between the pump and the discharge chamber, there is a recombinator supplied by Vacutech to catalyze atoms to parent molecules to prevent any chemical interaction of atoms with the oil in the vacuum pump. The atom densities are measured with movable catalytic probes supplied by Plasmadis (Ljubljana, Slovenia). The probe tip is moved along the discharge tube from the coil toward the flange next to the pump, as shown in Fig. 4.31. The probe tip is never inside the coil, where intensive plasma appeared in the H-mode. The probe would melt immediately if placed in a powerful plasma inside the coil according to the considerations presented in Figs. 4.10 or 4.11. The recombination coefficients for the catalytic tips are 0.12, 0.066, and 0.50 for O, H, and N atoms, respectively. An optical spectrometer is connected to a movable optical fiber for measuring characteristic spectra in the range between approximately 200 and 1100 nm. The vacuum system is often vented to an ambient atmosphere and never baked; therefore water is adsorbed on the inner surfaces.

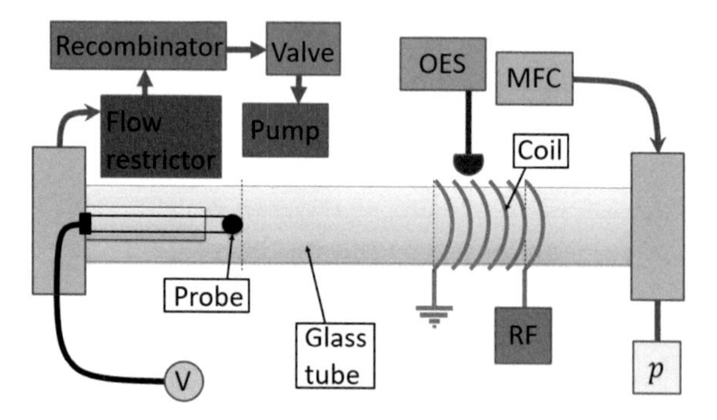

Figure 4.31 Schematic of the experimental system.

4.3.1 Oxygen Atoms

Oxygen of commercial purity 99.9% was introduced into the experimental system presented schematically in Fig. 4.32. A photo of the discharge tube at a discharge power of 100 W and a pressure of 20 Pa is shown in Fig. 4.33. The photo reveals three distinguished regimes: (i) plasma of a rather intensive radiation, which is limited to the coil volume, (ii) less luminous plasma between the coil and the grounded flange on the right, and (iii) very weak plasma in the volume between the coil and the flange on the left. The RF power is effectively transferred into the gaseous plasma in regime 1. This is due to the capacitive coupling between the coil and the gaseous plasma inside the coil. The electrons are accelerated in the sheath next to the coil between the glass tube and the plasma inside the coil. Once the fast electrons enter the plasma, they are quickly thermalized on elastic collisions with plasma electrons. Such constant supply of energetic electrons (the voltage across the sheath is hundreds of volts) from the sheath to the plasma inside the coil makes the electron temperature rather high in regime 1. The glowing plasma, however, does not terminate at the end of the coil but expands 10 cm away, toward the grounded flange on the right. The reason for such expansion is the capacitive coupling between the coil and the flange. The reason for the appearance of a rather uniform plasma in regime 2 is simple: the capacitive component of an impedance across the glass wall next to the grounded part of the coil is larger than the resistance of the plasma of a length of approximately 10 cm. As explained in Section 4.2, the electrons have large mobility; therefore the impedance of the plasma itself (without sheaths) is easily approximated by a pure resistive impedance, which is usually much lower than other impedances involved in the electrical circuit. In fact, the plasma of regime 2 does not terminate at the grounded flange but expands into the entire metallic tube (not shown in Fig. 4.33). This is called the "hollow electrode effect." The reason for such an expansion is the fact that the resistance of the plasma is less than the capacitive component of the sheath impedance. It is favorable to have a large "capacitor" (i.e., the surface of the sheath next to the grounded flange) and larger resistance in series than vice versa.

The same considerations apply for the flange on the left, except that the length between the coil and this flange is larger. Because a larger length results in a larger resistance of the capacitive coupling of the asymmetric configuration as adopted in this work, the plasma is much more intensive to the right of the discharge tube than to the left. If the configuration was symmetric, regime 2 would have also been observed on the left side of the discharge tube, provided the pressure is low enough to allow for a rather high mobility of electrons. At higher pressures, a capacitive discharge is established on one side only. From these considerations, one can deduce that the plasma would be limited to regime 1 only if the capacitive impedance across the glass wall next to the coil is lower than the resistance of the plasma in regimes 1 and 2, that is, if the distance between the coil and any grounded flange and/or the pressure is large.

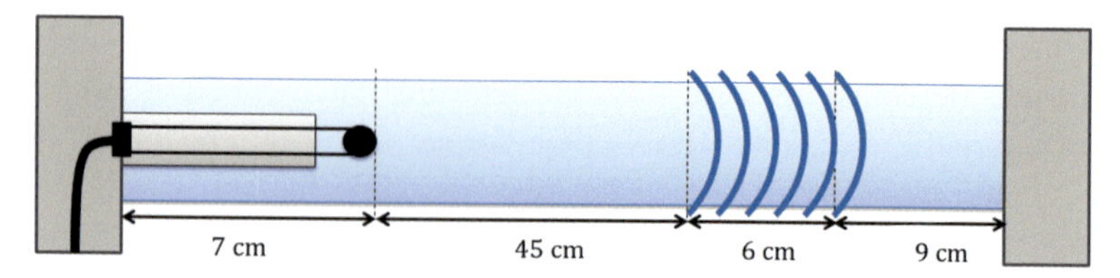

Figure 4.32 Details of the configuration of the discharge tube for measuring the density of O atoms.

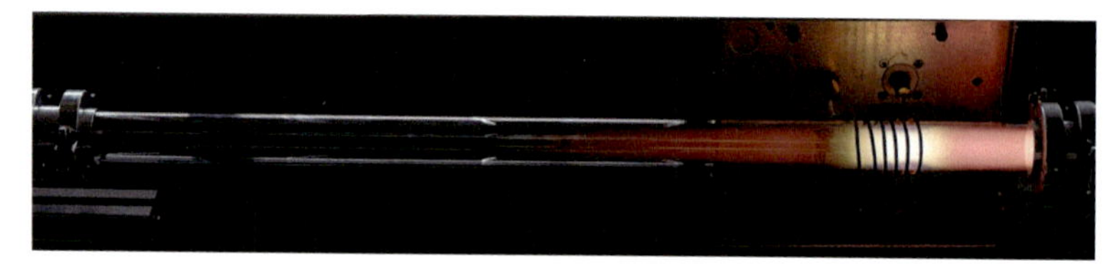

Figure 4.33 Photo of the discharge tube when oxygen plasma is in the E-mode.

Figure 4.34 shows an image of oxygen plasma at 500 W (high power) and a pressure of 20 Pa when it is in the H-mode. The plasma luminosity in the volume inside the coil is much larger than in the case of E-mode, which is presented in Fig. 4.33. The very luminous plasma indicates a high electron density. Therefore, the coupling as shown in Fig. 4.24 prevails. The coupling is thus almost entirely inductive in Fig. 4.34. The voltage is somewhat smaller in the H-mode and the current larger. However, they do not differ dramatically between the E- and H-modes [26]. The main difference is in the phase shift between the voltage and the current. The dissipated RF power is a product of voltage, current, and phase shift, that is

$$P_{\text{plasma}} = U\,I\cos(\varphi). \tag{4.13}$$

In the E-mode, the $\cos(\varphi)$ is small, whereas in the H-mode, it is large. The plasma between the coil and flanges is not observed in Fig. 4.34 due to the large luminosity inside the coil, but it may occur because of the capacitive coupling between the coil and the flanges. The impedance of the capacitive coupling is large compared to the inductive impedance. Therefore, any effects due to the capacitive coupling are marginal when the plasma is in the H-mode.

Figure 4.34 Photo of the discharge tube when oxygen plasma is in the H-mode.

The optical spectra from the plasma sustained in the E- and H-modes are shown in Fig. 4.35. The spectra in both modes are similar—the most intensive radiation arises from transitions in the red part of the spectra at 777 nm and near infrared at 845 nm. The predominant line at 777 nm gives oxygen plasma its typical red color. The intensity of the radiation in the H-mode is roughly 3 orders of magnitude larger than that in the E-mode. The differences in intensities are explained by differences in the electron density and thus predominant coupling. Another spectral feature observed in the lower spectrum of Fig. 4.35 is attributed to the radiative transition of ionized oxygen molecules [27]. Although the height of the atomic line at 777 nm is much higher than the spectral features arising from molecular transitions in the broad range of around 600 nm, the oxygen plasma in the E-mode inside the coil does not assume red color but rather white. The reason for this impression is the fact that the molecular transitions are much broader than atomic and, therefore, the integral radiations at around 600 nm and 777 nm are comparable. The spectrum acquired from the plasma in the H-mode is free from molecular transitions. Such a spectrum indicates a very high dissociation fraction of oxygen in the H-mode.

The density of the oxygen atoms in the discharge tube shown in Fig. 4.32 obviously depends on the following parameters: the position inside the tube, the flow of gas through the tube, the pressure, and the power. The number of measurements for complete characterization is therefore too large, and only representative results are presented in the following text. The power indicated in all figures is the forward power as indicated by the power meter attached to the RF generator unless stated otherwise. It should be stressed again that the forward power is never equal to the power absorbed by the gaseous plasma. The forward power is always larger, and the real power absorbed by the plasma can be only a fraction of the forward power, depending on the coupling efficiency.

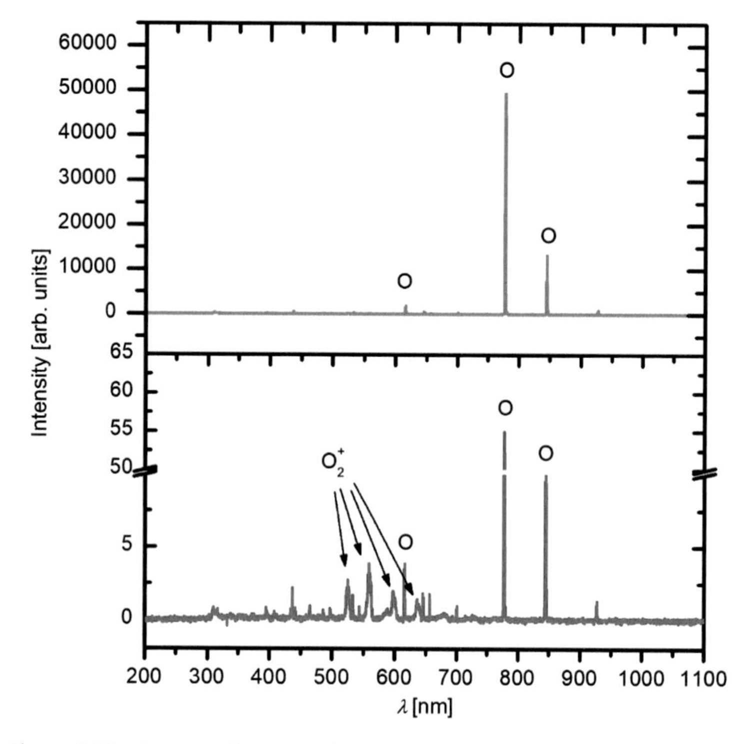

Figure 4.35 Spectra of oxygen plasma in the H-mode (upper curve) and the E-mode (lower curve).

The probe was fixed at the position indicated in Fig. 4.32, and the O atom density was measured at various pressures and discharge powers. Figure 4.36 represents the O atom density 7 cm away from the flange on the left in Fig. 4.32. The measurements were performed as follows: The values of the oxygen pressure and the forward RF power were preselected. The smallest pressure was set, and the forward power, supplied by the RF generator, was varied while the density of the O atoms was measured. Then, the pressure was changed and the measurements were repeated at the same forward

RF powers. The procedure was repeated until the entire range of pressures was characterized.

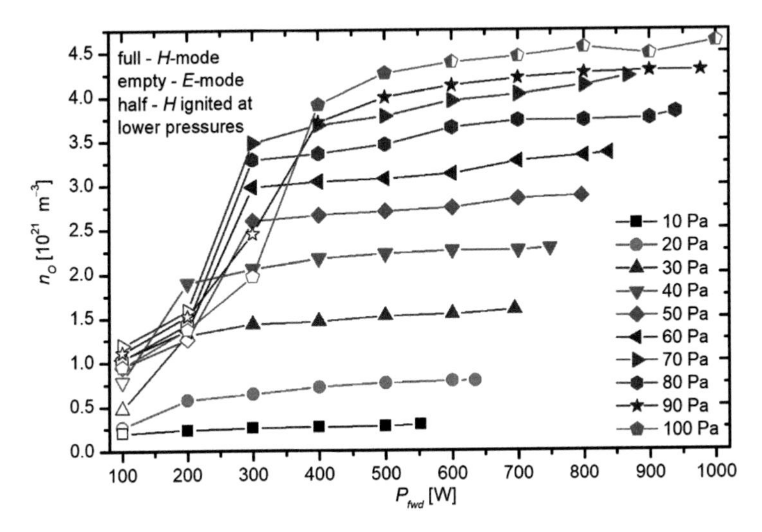

Figure 4.36 O atom density versus the forward power of the RF generator at different pressures inside the discharge chamber at the probe position 45 cm left from the coil. The setup is as in Fig. 4.32.

Figure 4.36 reveals the increase in the O atom density with increasing power. The effect is explained by more power absorbed in the plasma at increasing forward power. The density, however, does not increase monotonously with increasing power, but there is an instant increase that occurs at a certain forward power. The instant increase in the O atom density is explained by the transformation of the discharge coupling from the E- to the H-mode. As long as the discharge is in the E-mode, the density is up to approximately 1×10^{21} m^{-3}. The density in the E-mode depends on the forward power only for pressures up to approximately 30 Pa and remains fairly constant (rather independent from pressure) thereafter. Obviously, the density of O atoms in the E-mode cannot be very high because of poor transfer of the available power for heating the plasma in the E-mode. When the discharge is in the H-mode, the O atom density increases monotonously with increasing power. This effect is explained by the fact that a variable fraction of the forward RF power is actually absorbed by the plasma. In the E-mode, the fraction of the power absorbed by the plasma using the setup as in Fig. 4.32 is rather small. Therefore, any increase in the forward power does not reflect in the increase in the absorbed power and thus in the density of atoms. This effect is often overlooked in scientific literature.

The experimental system is equipped with a meter of reflected power. Because the accuracy of measuring the reflected power is not the best, it may be more appropriate to present the O atom density versus the difference between the forward and reflected powers. One such graph is shown in Fig. 4.37. In this figure, the curves are monotonous, so such a presentation of measurements is more appropriate. It appears rarely in scientific literature, though.

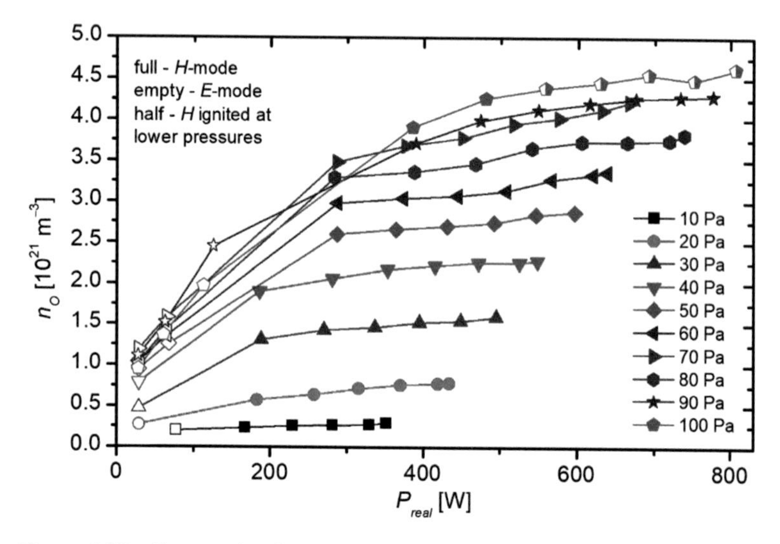

Figure 4.37 O atom density versus the difference between the forward and reflected powers at different pressures inside the discharge chamber at the probe position 45 cm left from the coil. The setup is as in Fig. 4.32.

The difference between the forward and reflected powers is a better parameter than the forward power itself. However, it still cannot be equalized with the power absorbed by the plasma. Such measurements have rarely been reported because they are difficult to realize. An interesting feature is observed in Fig. 4.37: the measured points are not equidistant, but there is a gap between the E- and H-modes. For example, the O atom density at the pressure of 100 Pa was determined for powers up to approximately 100

W and above 400 W. In between, it was impossible to sustain the gaseous plasma using the configuration in Fig. 4.31. The powers up to 100 W are attributed to a discharge in the E-mode and above 200 W to the H-mode. Between the modes, there is a power gap that is sometimes called the "forbidden zone." This effect occurs due to different discharge couplings. As the forward power increases in the E-mode, the reflected power increases too. Therefore, the increase in the difference in powers is not linear with increasing forward power. The physical explanation of the forbidden zone is as follows: As the power on the coil increases in the E-mode, the voltage increases, too. Once a certain power is reached, the voltage in the E-mode is equal to the maximal voltage of the power supply. Although an attempt is made to increase the power to reach the value that was set, the forward power cannot be increased any more in the E-mode because the maximal available generator voltage has been reached.

The situation is completely different in the H-mode. The coupling is predominantly inductive, as shown schematically in Fig. 4.24. For inductive coupling, a lower voltage is necessary to sustain a gaseous plasma. Therefore, the power supply does not reach the maximal available voltage. The necessary condition for the plasma in the H-mode is a large conductivity, that is, a high density of electrons, which cannot be obtained at a low discharge power. When the pressure is 100 Pa, the plasma in the H-mode is only sustained at powers over approximately 400 W. This is a simple but useful explanation of the forbidden zone of powers as revealed from Fig. 4.37.

Another law is evident from Fig. 4.37: the power gap (the difference between the minimal power achievable in the H-mode and the maximal power achievable in the E-mode) increases with increasing pressure: at 10 Pa, it is roughly 40 W (Fig. 4.38); at 20 Pa, it is already approximately 60 W; whereas at 100 Pa, it is as much as 250 W. The width of the power gap (forbidden zone) increases monotonously with increasing pressure. Figure 4.38 represents this effect. Here, it should be stressed that the curve in Fig. 4.38 is valid for the system shown in detail in this chapter only. If the configuration is different, the width of the power gap is different.

Figure 4.38 also reveals an important rule: plasma can be sustained in the H-mode at elevated pressures only if the RF generator is powerful enough and if the matching network allows for the efficient transfer of the available generator power into gaseous plasma. As the pressure decreases, the power gap decreases as well. In fact, the transition between the E- and H-modes is not abrupt at pressures below approximately 1 Pa for the configuration presented in this section. The effect is explained by the decreasing frequency of electron collisions and thus an increase in the mean free path of the electrons with decreasing pressure. As a consequence, the electrons have higher mobility at low pressures and thus the ability to sustain a dense plasma at a relatively low power at low pressures.

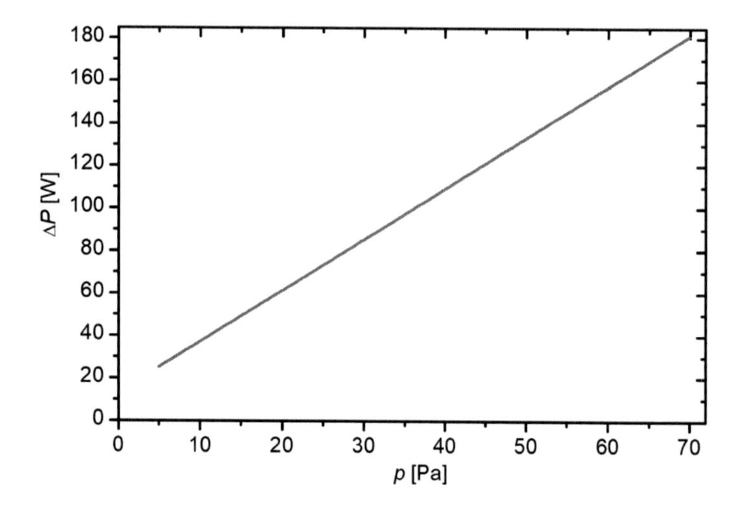

Figure 4.38 Width of the power gap between the E- and H-modes (forbidden zone) versus the oxygen pressure in the discharge tube of the configuration shown in Fig. 4.32.

Usually, it is useful to present the density of O atoms versus the pressure with power as a parameter, as in Fig. 4.39. One can observe an interesting effect: all curves (except the lowest) in Fig. 4.39 follow the general curve and deviate at a certain pressure, which depends on the preset generator power. This observation is useful when one designs the discharge configuration. Obviously, a large generator power is unnecessary at a low pressure. The effect has been already explained, but it is useful to recall it: as long as the plasma is in the

H-mode, the electron density is very high in the plasma within the coil. As a consequence, the molecules are highly dissociated already at a moderate power. So, any increase in the discharge power does not result in any dissociation but in other effects, such as increased electron density and more energy dissipation for phenomena other than dissociation. The deviation of the atom density from the general curve presented in Fig. 4.39 obviously occurs when the discharge mode is transformed from the H- to the E-mode. Here, it is useful to mention that if the power is increased while the plasma is already ignited, the transition from the E- to the H-mode does not appear at the same forward power as from the H- to the E-mode if the power is decreased. Therefore, a hysteresis is observed. The hysteresis is due to the fact that the power needed for ignition of the H-mode is larger than the power needed for sustaining the H-mode at a given pressure. As the pressure increases, the ignition power increases too. Therefore, at an elevated pressure, the available generator power is not sufficient to ignite the H-mode but enough to sustain it. A practical consequence of this effect is that one can first ignite the H-mode at a rather low pressure at a given generator power and then increase the pressure to sustain the H-mode at an elevated pressure. This praxis was adopted in some measurements shown in Fig. 4.39.

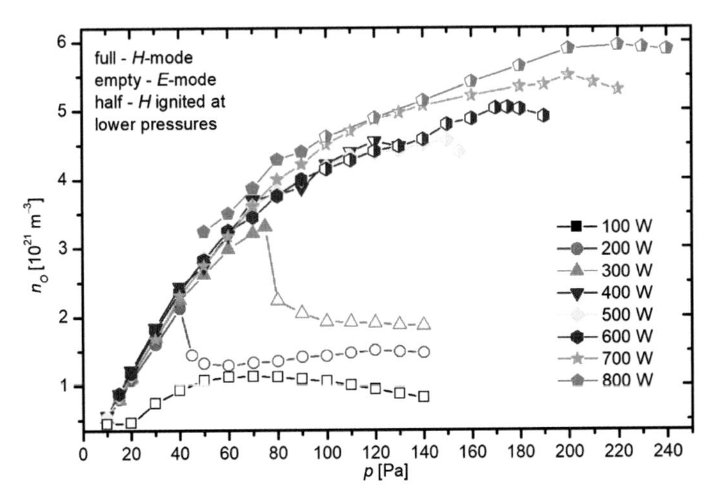

Figure 4.39 O atom density at the probe position 45 cm away from the RF coil, toward the flange on the left in Fig. 4.32. The parameter is the forward power.

Another useful parameter is the dissociation fraction, which is defined as the ratio between the atom density and twice the density of the molecules in the discharge chamber before the discharge is ignited. The dissociation fraction for the experimental conditions adopted for measuring the O atom densities as in Fig. 4.39 is shown in Fig. 4.40. The dissociation fraction depends largely on the discharge mode: in the E-mode, it is several times lower than in the H-mode. The difference is explained by the more efficient dissociation of molecules when the discharge is in the H-mode. As already stated, the electron density in the H-mode is much larger than in the E-mode. A similar behavior of the oxygen dissociation fraction is observed as for the case of O atom density: there is a general curve in the H-mode. As long as the plasma is in the H-mode, the dissociation fraction does not depend much on the discharge power because the dissociation is efficient enough already at moderate powers as long as the pressure is not too high.

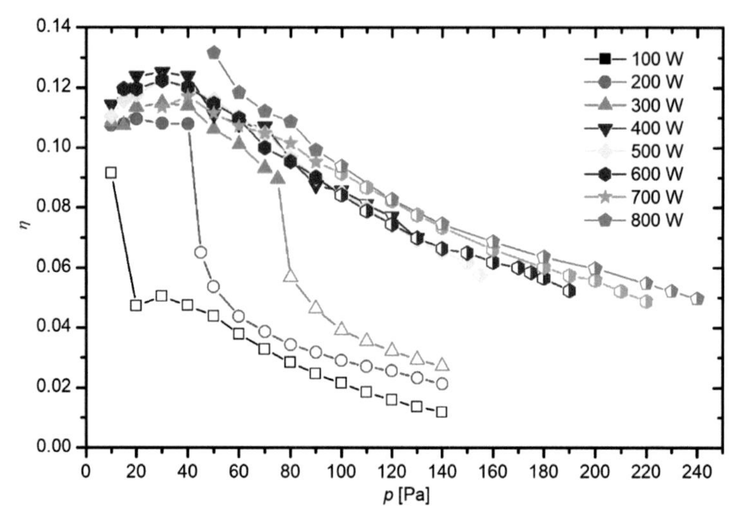

Figure 4.40 Dissociation fraction of oxygen molecules 45 cm away from the RF coil, toward the flange on the left in Fig. 4.32. The parameter is the forward power.

The density of the oxygen atoms at a given position in the discharge tube does not depend only on the pressure and power but also on the time the gas spends on the way from the intensive plasma within the coil to the position of the probe. Figure 4.41 reveals this effect. The curves reveal a monotonous increase in the O atom density with an increasing flow rate. This observation indicates that the residence time of the gas flowing through the intensive plasma inside the coil does not affect the density of the atoms downstream. If the residence time was too short to ensure appropriate dissociation of oxygen molecules, the curves would have decreased with increasing flow at large flows. Theoretically, there should be a maximum in the curves presented in Fig. 4.41: when the flow is zero, the O atom density should be very low due to poor migration (by diffusion) of atoms from the plasma inside the coil (where they are formed by an electron-impact dissociation of parent molecules) to the position of the probe. When the flow is infinitely large, the residence time should approach zero, also resulting in a poor atom density at the probe position because the electrons would miss numerous molecules. In practice, however, the flow is limited due to the limited gas drift (the theoretically maximum drift velocity is equal to the sound velocity, as explained in Section 4.1) and due to a finite pumping speed. The latter limitation prevails at experimental conditions adopted in this study. The fact that no extreme is observed in the curves in Fig. 4.41 therefore clearly indicates that the residence time of the gas in the plasma inside the coil does not influence the O atom density in the experimental system adopted in this work.

The monotonous increase in the O atom density with an increasing gas flow in Fig. 4.41 also indicates that the dissociation of oxygen molecules by plasma electrons in region 3 of Fig. 4.33 is not efficient enough to replace the loss of atoms on the way from the coil to the probe position. If it was sufficient, the curves in Fig. 4.41 would have been constant. Region 3 in Fig. 4.33 indicates rather poor capacitive coupling between the coil and the grounded flange on the left—see Figs. 4.33 and 4.34. Obviously, the production of atoms in such a weak discharge is insufficient for a reasonable atom density. The loss of atoms at this pressure (50 Pa) is almost exclusively by heterogeneous surface recombination. To minimize this loss, the gas flow should be rather large. Although the recombination coefficient for oxygen atoms on the surface of borosilicate glass is low at about 10^{-3}, the loss of atoms is not negligible.

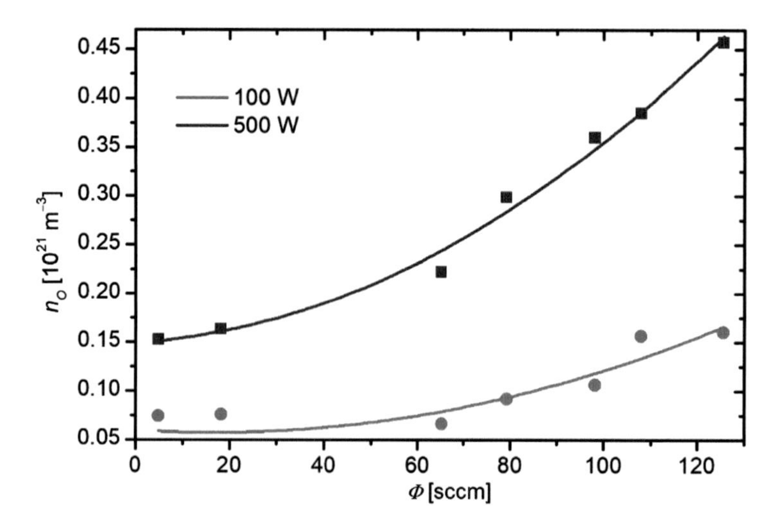

Figure 4.41 Density of O atoms 45 cm away from the RF coil, toward the flange on the left in Fig. 4.32, versus the flow of the gas through the discharge tube at a pressure of 50 Pa when the plasma is in the E-mode (lower curve) and H-mode (upper curve).

The effect of decreasing atom density due to surface recombination with parent molecules can be monitored using a movable probe. Figure 4.42 reveals the dissociation fraction versus the probe position and other fixed parameters, that is, the gas pressure of either 20 or 50 Pa and the forward power of either 100 or 500 W. The dissociation fraction slowly decreases with increasing distance from the coil, which is explained by weak surface recombination because of the small recombination coefficient. The curves are linear except close to the coil when the plasma is in the H-mode (at 500 W). The deviation close to the coil is explained by the fact that the intensive plasma causes a very high dissociation fraction in the plasma. Therefore, the O atom density in the plasma within the coil is much larger than at any probe position adopted in this experiment. The dissociation fraction as shown in Fig. 4.42 is moderate—only 20% at the closest position to the coil, which suggests a discrepancy in the claim that the oxygen molecules are almost fully dissociated in a dense plasma inside the coil. The virtual discrepancy is explained by thermal effects: although the discharge tube is forced-air cooled, the inner surface of the glass tube warms up when the plasma is in the H-mode. As a consequence, the O atoms are lost more efficiently close to the coil because the recombination coefficient increases with temperature. At a distance of 15 cm from the coil, the discharge tube is cold (at room temperature), irrespective of the discharge power. Therefore, the atom density slowly decreases with increasing distance from the coil in this part of the tube.

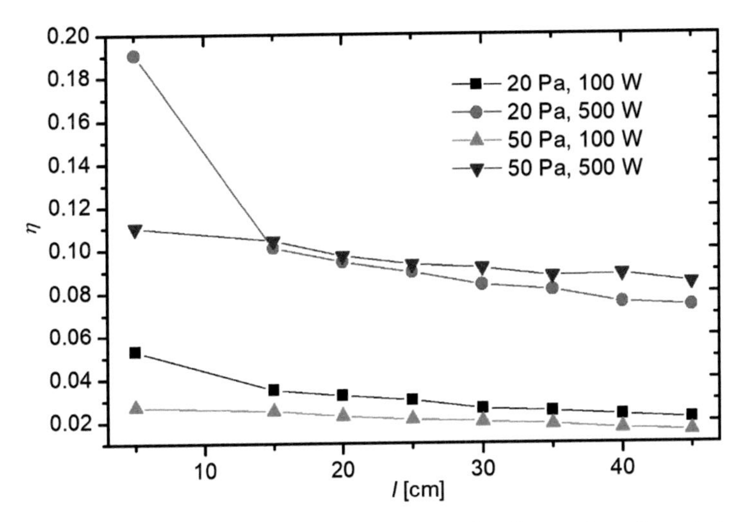

Figure 4.42 Oxygen dissociation fraction versus the distance from the coil at a fixed flow and two different pressures and forward powers.

4.3.2 Hydrogen Atoms

The dissociation energy of hydrogen molecules is lower than that of oxygen. Therefore, one would expect larger dissociation fractions at the same discharge condition. Such a simple consideration, however, is not justified for the experimental setup adopted in this research. As mentioned earlier, the recombination coefficient for hydrogen atoms on the surface of borosilicate glass is larger than that for oxygen and the coefficient increases significantly with increasing temperature. These effects influence the density of H atoms in the experimental system shown in Fig. 4.31. The configuration of Fig. 4.32, which proved very useful for oxygen, failed for the case of hydrogen. A reason for this is the fact that at the probe position 45 cm away from the coil, the atom density was close to the detection limit of the probe when the discharge was in the E-mode. The system was thus slightly modified to enable comprehensive characterization of the RF source of H atoms. The details of the discharge tube adopted for experiments with hydrogen are shown in Fig. 4.43.

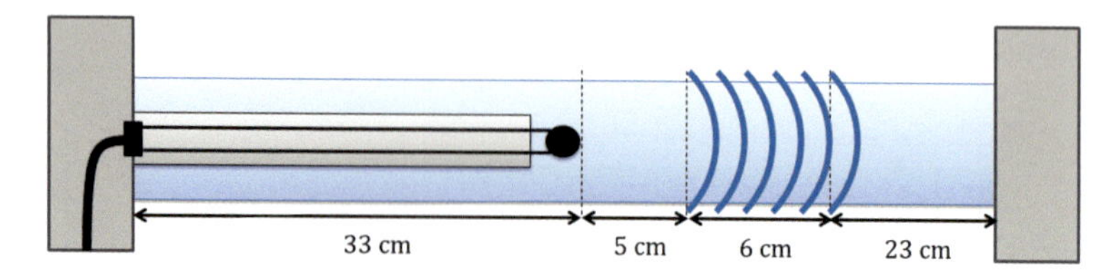

Figure 4.43 Details of the configuration of the discharge tube for measuring the density of H atoms.

A photo of hydrogen plasma in the E-mode is shown in Fig. 4.44. The luminous plasma expands throughout the discharge tube and is more luminous inside the coil. Almost equal plasma luminosity in the tube both left and right from the coil indicates almost equal capacitive coupling between the coil and the grounded flanges. The difference between hydrogen and oxygen plasmas is explained by different positions of the coil. In the case of oxygen (Fig. 4.33), the coil was placed highly asymmetrically, whereas in the case of hydrogen (Fig. 4.43), an almost symmetric geometry was chosen. The more luminous plasma inside the coil is explained by the larger electron temperature due to the constant supply of high-energy electrons from the sheath between the plasma and the glass tube backed by the powered coil, as explained in detail for the case of oxygen plasma.

Figure 4.44 Photo of the discharge tube when hydrogen plasma is in the E-mode, that is, at 20 Pa and 150 W forward power.

Figure 4.45 shows a photo of hydrogen plasma in the H-mode. The luminous plasma is now concentrated in the volume inside the coil, and only very weak plasma expands over the rest of the discharge tube. The reasons are the same as explained for oxygen plasma. The interface between the luminous plasma within the coil and the dim plasma elsewhere is rather sharp. In the volume of the luminous plasma, the coupling is the same as shown schematically in Fig. 4.24. The dense plasma expands a few centimeters outside the coil, where the density of electrons is still high enough to ensure effective inductive coupling.

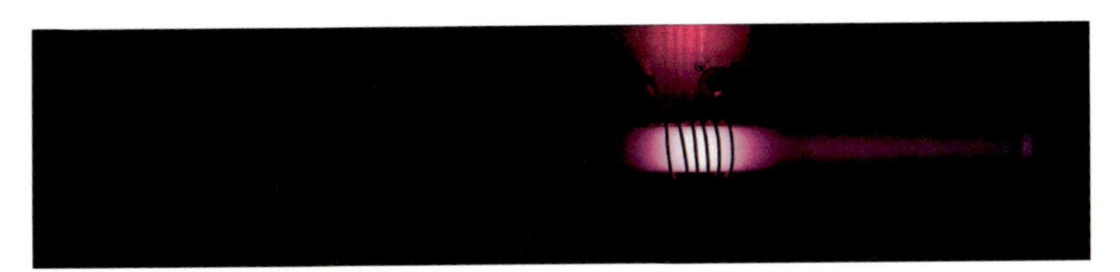

Figure 4.45 Photo of the discharge tube when hydrogen plasma is in the H-mode, that is, at 20 Pa and 500 W forward power.

The spectra of hydrogen plasma in the E-mode (photo in Fig. 4.44) and the H-mode (Fig. 4.45) are shown in Fig. 4.46. In both modes the atomic transitions indicated by sharp lines prevail. The atomic lines indicate transitions from a higher to the first excited state. The transitions to the ground state (although more intensive than those to the first excited states) are not observed because they appear in the far-UV range of the spectra, not detectable by the optical spectrometer. Instead, one can observe other features when the plasma is in the E-mode (lower curve in Fig. 4.46). There are bands peaking at approximately 600 nm, which correspond to neutral molecule transitions (Fulcher band). There is also a broad continuum in the near-UV range of the spectrum. The continuum arises from relaxation of triplet molecules from the excitation level at about 12 eV to a lower state that dissociates immediately. Because the lower state is not bonded (molecules in this state dissociate immediately), the radiation from this triplet state causes a continuum. The continuum observed in the lower curve of Fig. 4.46 terminates at the wavelength of approximately 300 nm because of the fact that the borosilicate glass is not transparent for radiations of shorter wavelengths. The spectrum of hydrogen plasma in the H-mode (upper curve in Fig. 4.46) reveals only atomic transitions. This observation indicates that the dissociation fraction of hydrogen molecules when the plasma is in the H-mode is very high, the same as for oxygen plasma.

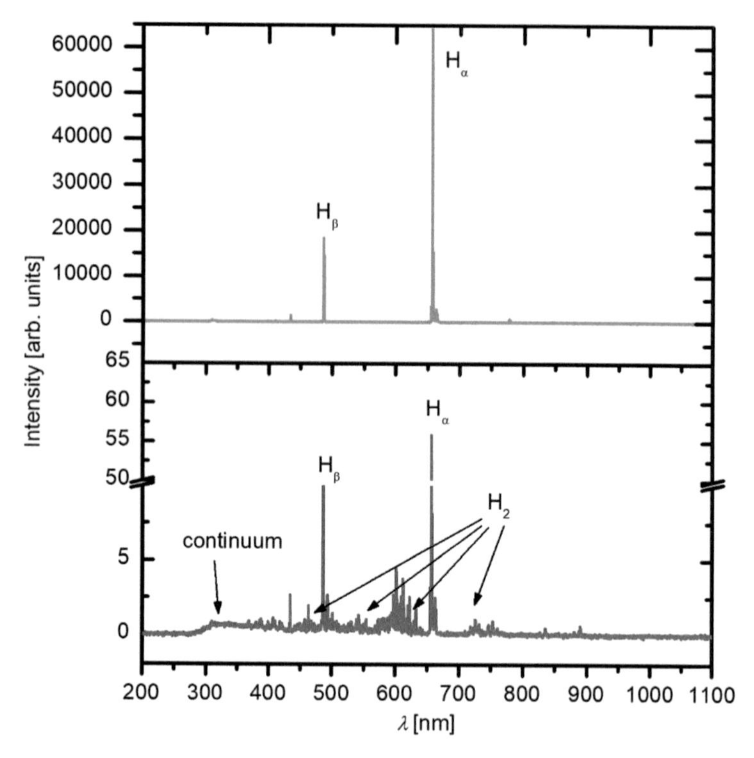

Figure 4.46 Spectra of hydrogen plasma in the E-mode (lower curve) and the H-mode (upper curve).

The density of H atoms in the configuration presented in Fig. 4.43 was measured with the catalytic probe, whose tip was placed 5 cm from the coil in the direction of the gas flow (see Fig. 4.43). The measurements were performed using the same procedure as for oxygen. The H atom density versus the forward power of the RF generator is plotted in Fig. 4.47. As in the case of oxygen, there is an abrupt increase in the H atom density at a certain power, which depends on the pressure. At a pressure of 15 Pa, an abrupt increase in the H atom density appears at the forward power of approximately 250 W, whereas at 50 Pa, it appears at approximately 400 W. As already stressed, the forward power is not an indicator of the power absorbed in the plasma. To this end, it is better to present the H atom density versus the difference between the forward and reflected powers as shown in Fig. 4.48. The curves are now rather monotonous: the H atom density increases fairly linearly with increasing power up to the highest available powers. This effect is somewhat different than for the case of oxygen, where some sort of saturation was observed at elevated powers (see Fig. 4.37). The discrepancy is explained by different distances between the luminous plasma inside the coil and the probe position—the probe was closer in the case of hydrogen plasma. As shown in Fig. 4.42, the H atom density increases rapidly with decreasing distance close to the coil.

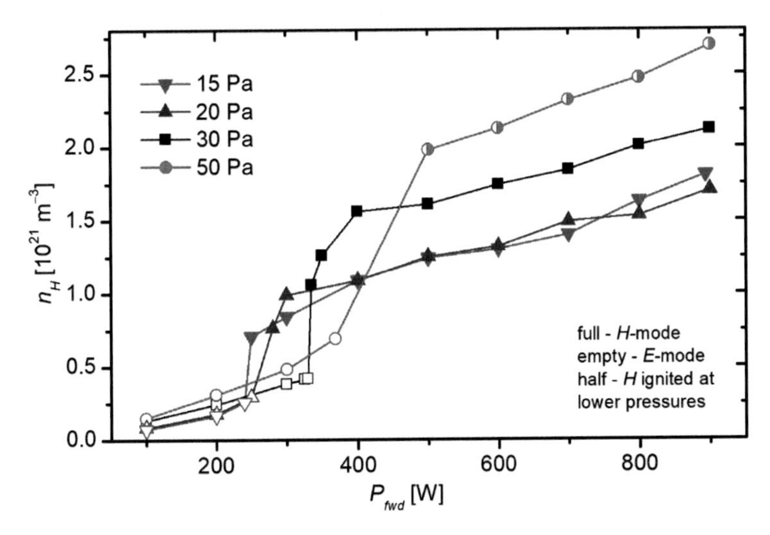

Figure 4.47 H atom density versus the forward power of the RF generator at different pressures inside the discharge chamber at the probe position 5 cm left from the coil. The setup is as in Fig. 4.43.

Similar to the case of oxygen (Figs. 4.37 and 4.38), a forbidden zone of powers is also observed for hydrogen, as shown in Fig. 4.48. The range of forbidden powers increases with increasing pressure. The reason is already explained for the case of oxygen: insufficient voltage of the power supply when the discharge is in the predominant capacitive mode. The width of the power gap (the forbidden zone), that is, the difference between the minimal power for sustaining the plasma in the H-mode and the maximum power achievable for sustaining the plasma in the E-mode is similar to the width observed in the oxygen plasma (Fig. 4.38). The width in the case of the hydrogen plasma at experimental conditions adopted in this work is shown in Fig. 4.49.

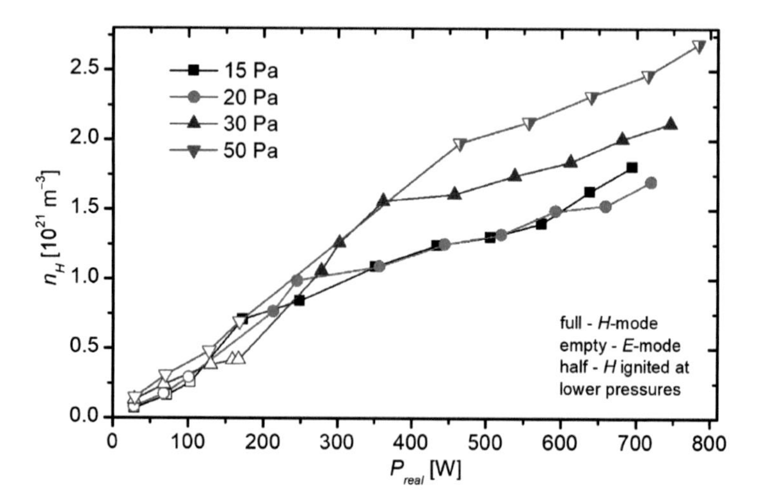

Figure 4.48 H atom density versus the difference between forward and reflected powers at different pressures inside the discharge chamber at the probe position 5 cm left from the coil. The setup is as in Fig. 4.43.

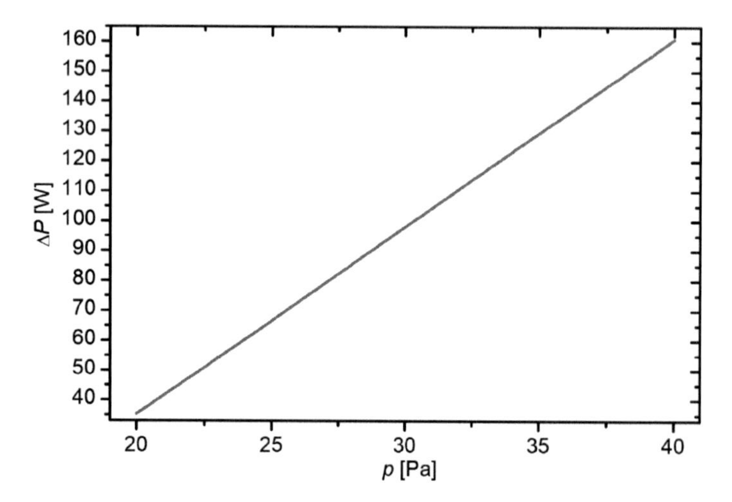

Figure 4.49 Width of the power gap between the E- and H-modes (forbidden zone) versus the hydrogen pressure in the discharge tube of the configuration shown in Fig. 4.43.

The density of hydrogen atoms at the position 5 cm from the coil (see Fig. 4.43) versus the pressure is shown in Fig. 4.50. The forward power is the parameter. The H atom density was measured at the preset forward powers of 100, 500, and 700 W. When the forward power was 100 W the discharge was in the E-mode for the entire range of pressures, whereas for 500 and 700 W, it was in the H-mode. The hydrogen plasma in the H-mode does not ignite at our experimental conditions in the H-mode at pressures above 35 Pa. Therefore, it was ignited at a lower pressure and sustained at an elevated pressure. As mentioned earlier, this procedure was adopted to benefit from the fact that the power needed to sustain the plasma in the H-mode is lower than the power needed to ignite the plasma in the H-mode.

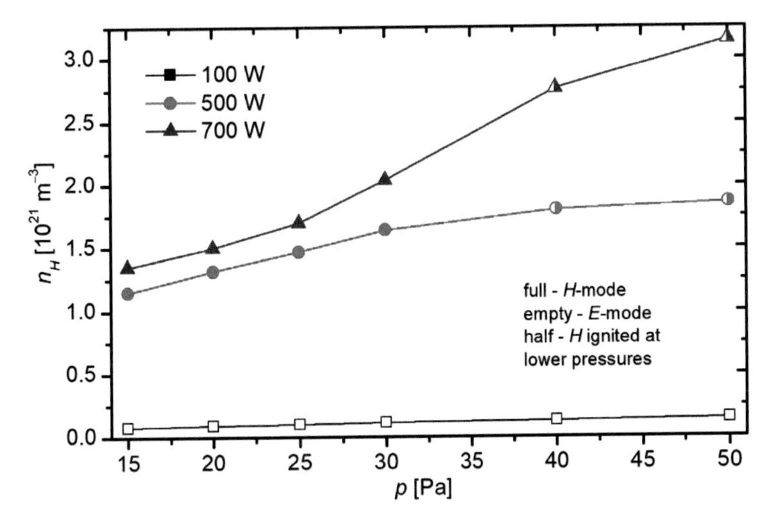

Figure 4.50 H atom density 5 cm away from the coil in Fig. 4.43. The parameter is the forward power.

Figure 4.50 reveals that the H atom density is similar to the O atom density (Fig. 4.39) as long as the hydrogen plasma is in the H-mode. In the E-mode, however, it is much lower. The lowest curve in Fig. 4.50 indicates a H atom density of around 10^{20} m^{-3}, which is much lower than that for oxygen. The dissociation energies are comparable (5.16 eV for an oxygen molecule and 4.52 eV for a hydrogen molecule [2]). Therefore, one would also expect comparable atom densities. The reason for the poor dissociation of hydrogen molecules 5 cm away from the coil is the more efficient surface recombination of H atoms. The borosilicate glass exhibits a moderate value of the recombination coefficient for H atoms (see

Table 4.3), which reflects in a poor dissociation fraction of hydrogen molecules (below 1%) when the plasma is in the E-mode. Such a weak discharge is therefore a useful source of O atoms, but for H atoms, one should use an H-mode plasma to create large densities of H atoms in a borosilicate discharge tube powered by an RF generator.

There is no "general curve" for the hydrogen plasma in the H-mode in Fig. 4.50, as observed for the oxygen plasma (Fig. 4.39), but the H atom density is much higher at 700 W than at 500 W. Obviously, a large amount of power even in the H-mode is necessary to achieve good dissociation of hydrogen molecules 5 cm away from the discharge tube. This effect can be attributed to the fact that the luminous hydrogen plasma in the H-mode (Fig. 4.45) stretches outside the coil and the expansion increases with increasing discharge power.

The results presented in Figs. 4.47–4.50 were obtained at a fully open flow restrictor. As the flow is decreased at a constant pressure in the discharge tube, the H atom density at the probe position 5 cm away from the coil decreases rapidly due to more extensive surface recombination and eventually reaches the detection limit of the probe; therefore, it is pointless to present results for hydrogen as are shown for oxygen in Fig. 4.41. More interesting are the results obtained with a movable probe, as shown in Fig. 4.51. The probe was moved from a distance 2.5 cm away from the coil to 35 cm, where the atom density reached the detection limit of the probe. A huge difference between the E- and H-modes is observed in Fig. 4.51 at close distances, but 15 cm away, the dissociation fractions are comparable. The dissociation fraction 2.5 cm away from the coil is more than 20%, indicating that inside the coil, it should approach 100%, which agrees with the upper curve of Fig. 4.46. When the plasma is in the H-mode, the hydrogen molecules are dissociated efficiently, but the dissociation fraction quickly decreases downstream from the luminous plasma. The reason for this undesired effect is extensive surface recombination of H atoms on the borosilicate glass. Unlike for the case of oxygen (Fig. 4.42), where a fairly linear decrease in atoms was observed along the discharge tube toward the pump, the decrease is extensive for the case of hydrogen. The lower curve in Fig. 4.51 might give an impression of linear dependence, but it is not so. To clarify this, the behavior of the hydrogen dissociation fraction versus the distance from the coil is plotted separately in Fig. 4.52.

In this figure, one can observe qualitatively the same behavior as the upper curve in Fig. 4.51, which is easily explained by the same recombination efficacy except that the initial H atom densities in the H- and E-modes are different. An important observation concerning hydrogen plasma is therefore the quick decay of the H atom density away from the plasma because of the extensive recombination of H atoms on the surface of the borosilicate glass.

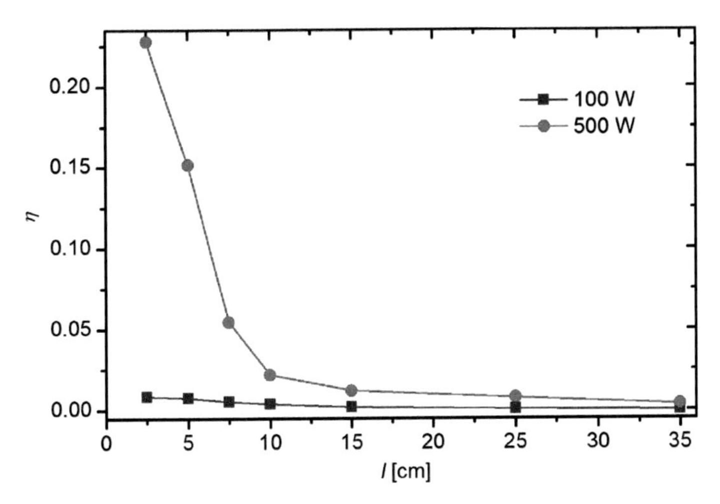

Figure 4.51 Hydrogen dissociation fraction versus the distance from the coil at the pressure of 30 Pa and forward powers of 100 W (plasma in the E-mode) and 500 W (plasma in the H-mode).

4.3.3 Nitrogen Atoms

Like O_2 and H_2, a nitrogen molecule consists of two atoms, but there are huge differences between these simple molecules. The dissociation energy of N_2 molecules is large, at about 9.8 eV. Therefore, one will immediately conclude that sustaining a plasma with a high dissociation fraction is a difficult task. However, nitrogen molecules are easily excited to high-energy vibrational levels, which are not quenched on superelastic collisions with species like H atoms. Therefore, the rotational temperature of N_2 molecules is usually measured in thousands of kelvins. There are several metastable states of N_2 molecules that accumulate the potential energy obtained because of plasma conditions.

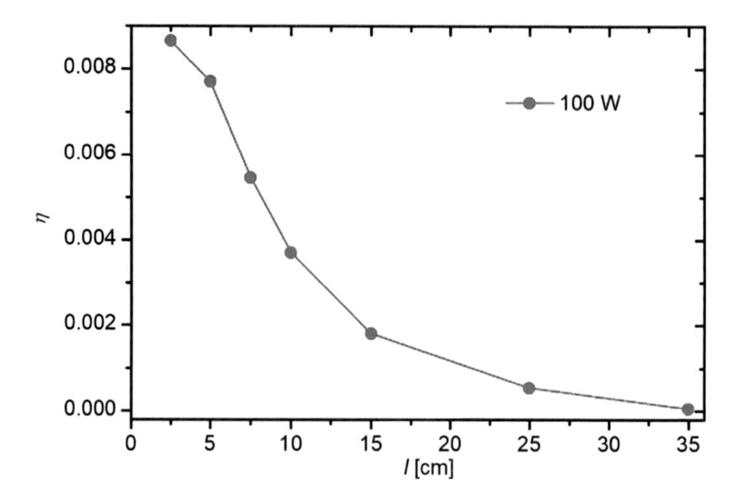

Figure 4.52 Hydrogen dissociation fraction versus the distance from the coil at the forward power of 100 W and the pressure of 30 Pa when the plasma is in the E-mode.

The density of nitrogen atoms was measured in the experimental system presented in Fig. 4.31. The RF coil was placed almost symmetrically in the discharge tube, as shown in Fig. 4.53. A photo of the discharge tube when the plasma is in the E-mode is shown in Fig. 4.54. The rather luminous plasma expands from the coil toward the grounded flanges, indicating good capacitive coupling, as shown schematically in Fig. 4.23. The plasma luminosity decays gradually from the coil toward the flanges, which is explained by a gradual decrease in the energy stored in the plasma particles (not only electrons) in the region inside the coil.

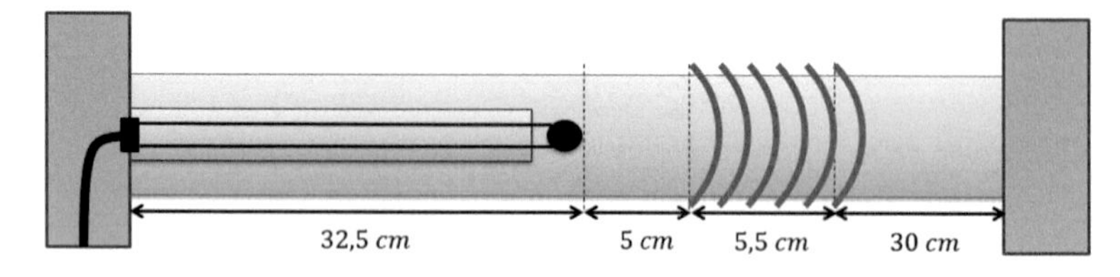

Figure 4.53 Details of the configuration of the discharge tube for measuring the density of N atoms.

A photo of nitrogen plasma in the H-mode is shown in Fig. 4.55. The luminous plasma is now concentrated within the coil, indicating the prevalence of inductive coupling, as shown schematically in Fig. 4.24. The much weaker plasma expands also to the grounded flanges because of the weak capacitive coupling. The effect has already been explained for oxygen and hydrogen plasmas.

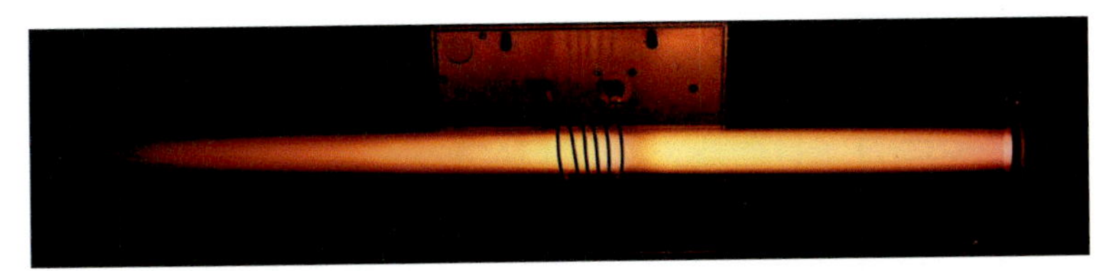

Figure 4.54 Photo of the discharge tube when nitrogen plasma is in the E-mode, that is, at 40 Pa and 200 W forward power.

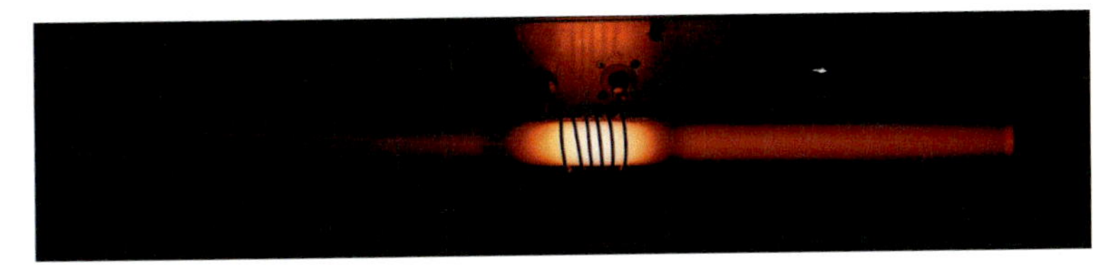

Figure 4.55 Photo of the discharge tube when nitrogen plasma is in the H-mode, that is, at 40 Pa and 500 W forward power.

The optical spectra for both E- and H-mode nitrogen plasmas are shown in Fig. 4.56. The lower spectrum was acquired in the E-mode and the upper one in the H-mode. The spectra are completely different from those acquired for oxygen and hydrogen plasmas. The radiation arising from atoms in the nitrogen plasma is negligible to that arising from neutral nitrogen molecules. The spectra shown in Fig. 4.56, therefore, qualitatively indicate a rather low dissociation fraction of nitrogen molecules, even when the discharge is in the H-mode, where the absorbed power is large. The main reason for such poor dissociation is a large dissociation energy and not the extensive surface recombination, which will become apparent from the measurements presented below.

The nitrogen spectra are rich in molecular transitions of neutral molecules. There are numerous vibrational bands. which are broadened by rotational bands, indicating that the rotational temperature is elevated, too, not only the vibrational one. In fact, nitrogen plasma at elevated discharge powers is never really cold but is still in a highly non-equilibrium state.

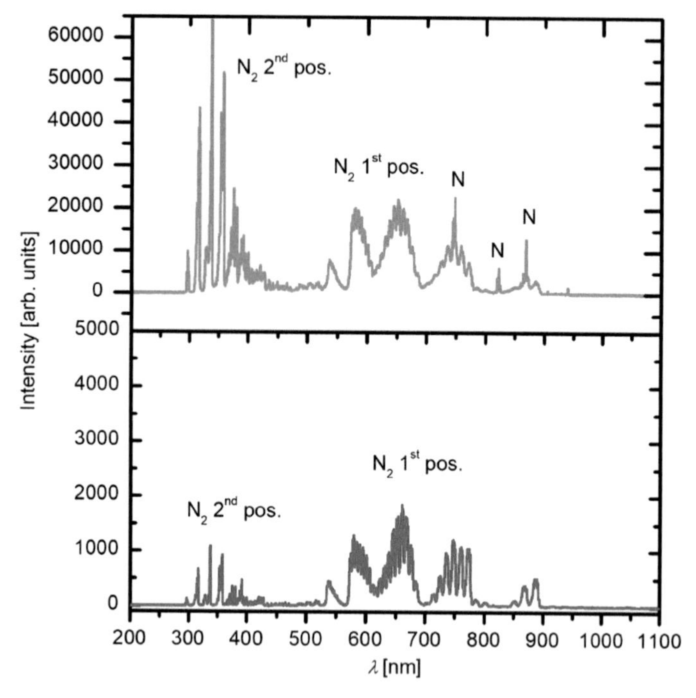

Figure 4.56 Spectra of nitrogen plasma in the E-mode (lower curve) and the H-mode (upper curve).

The density of N atoms was measured with a catalytic probe whose tip was placed 5 cm away from the RF coil in the direction of the gas flow, as shown in Fig. 4.53. The pressure of the nitrogen was set to a desired value, and the measurements were performed using the same routine as explained for oxygen and hydrogen. The results are plotted in Fig. 4.57. As for other gases, the same results are plotted versus the difference between the forward and reflected powers in Fig. 4.58. An important difference as compared to oxygen and hydrogen is the order of magnitude in the diagrams in Figs. 4.57 and 4.58. For oxygen and hydrogen, it was 10^{21} m^{-3}, whereas for nitrogen it is 10^{20} m^{-3}. The difference is explained by the very high dissociation energy of nitrogen molecule as compared to oxygen or hydrogen. Despite the high vibrational population as well as the presence of molecules excited to metastable states of a high excitation energy (over 6 eV), the dissociation fraction remains low—up to approximately 1% at the position 5 cm from the coil.

The forbidden zone (Fig. 4.59) is rather similar to that of oxygen (Fig. 4.38) and hydrogen (Fig. 4.49). Obviously, the forbidden zone is not affected much by the presence of atoms—rather it is correlated with impedances that are, in turn, correlated with charged particles in the gaseous plasma. The N atom density versus the pressure in the discharge chamber with forward power as the parameter is shown in Fig. 4.60.

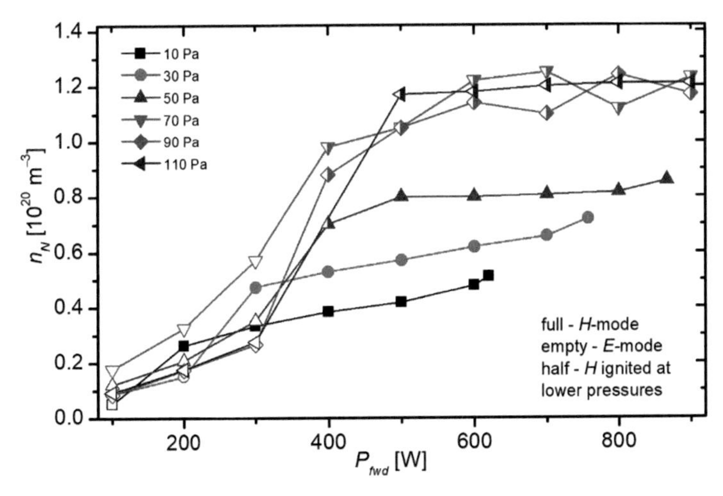

Figure 4.57 N atom density versus the forward power of the RF generator at different pressures inside the discharge chamber at the probe position 5 cm left from the coil. The setup is as in Fig. 4.53.

The N atom density was also determined with a movable probe, and the results are presented in Fig. 4.61. The N atom density decreases slowly with increasing distance from the coil, indicating behavior qualitatively similar to that observed for oxygen (Fig. 4.42). Such a slow decrease reveals a reasonably low loss of N atoms on the way from the powerful plasma inside the coil toward the flange in the direction of the gas flow. As for oxygen, the loss of nitrogen atoms is attributed predominantly to heterogeneous recombination on the surface of the borosilicate glass. The recombination coefficient for N atoms is thus similar to that of O atoms and definitely much smaller than for H atoms, for which a rapid decay of the H atoms was observed (Figs. 4.51 and 4.52). The somewhat larger values of the nitrogen dissociation fraction close to the coil when the plasma is in the H-mode is explained by the same arguments as for the oxygen plasma: deviations because of thermal effects. Such deviations are not observed as long as the nitrogen plasma is in the E-mode when the glass tube remains close to room temperature.

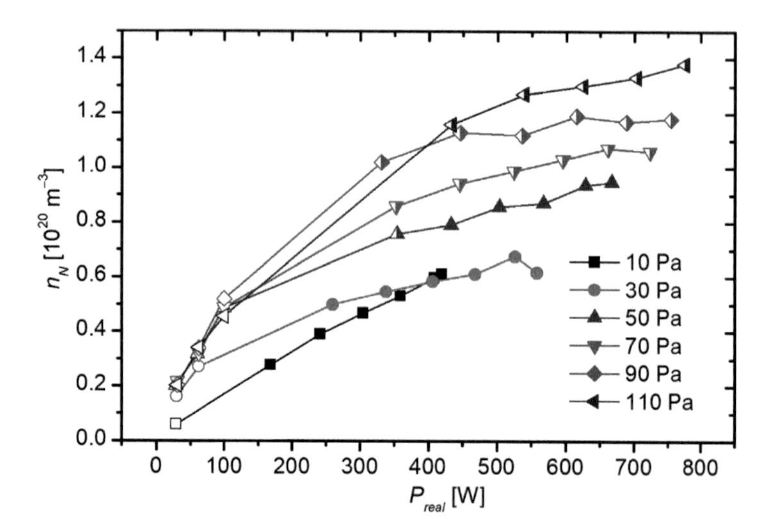

Figure 4.58 N atom density versus the difference between the forward and reflected powers at different pressures inside the discharge chamber at the probe position 5 cm left from the coil. The setup is as in Fig. 4.53.

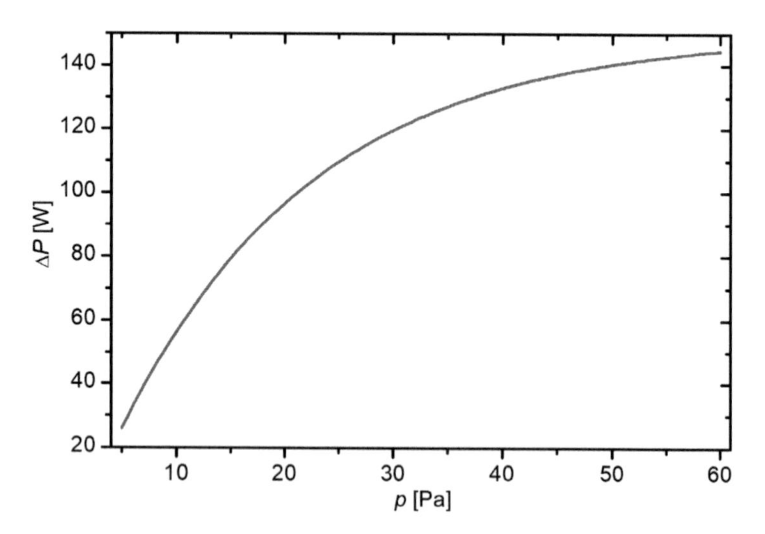

Figure 4.59 Width of the power gap between the E- and H-modes (forbidden zone) versus the nitrogen pressure in the discharge tube of the configuration shown in Fig. 4.53.

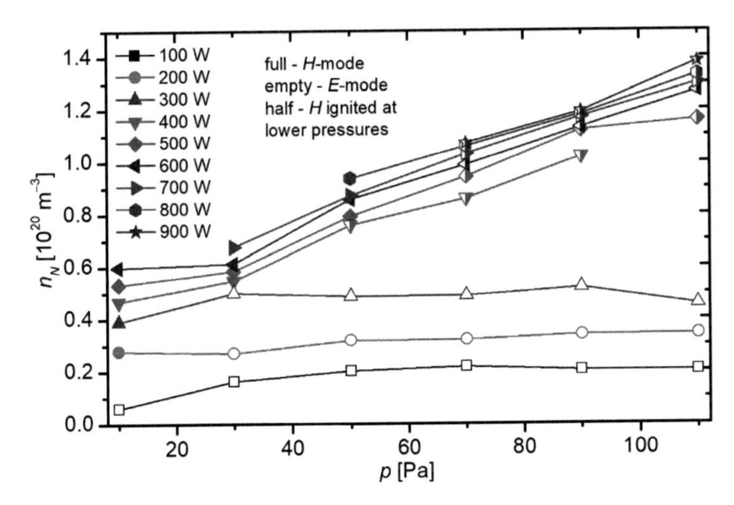

Figure 4.60 N atom density at the position 5 cm left from the coil as shown in Fig. 4.53. The parameter is the forward power.

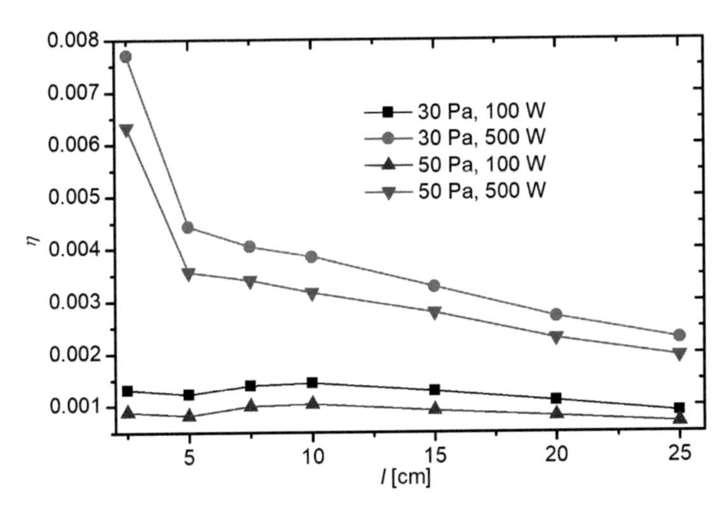

Figure 4.61 Nitrogen dissociation fraction versus the distance from the coil at pressures of 30 and 50 Pa and forward powers of 100 (plasma in the E-mode) and 500 W (H-mode).

4.4 Conclusions

Two-atom molecules, such as oxygen, hydrogen, and nitrogen, are effectively dissociated into parent atoms in a gaseous plasma sustained by electrodeless RF discharges in a dielectric tube. Although a coil connected to the RF generator via a matching network is applied, the predominant coupling of the plasma with the RF field is not always inductive. At the low power absorbed by a gaseous plasma, the predominant coupling is capacitive. As the power increases, the transformation from capacitive to inductive coupling occurs instantly and so does an increase in the atom density. The atom density in a discharge tube in the predominantly inductively coupled power is always larger than in the capacitively coupled power because of a higher electron density in such plasma. The density of atoms away from the inductive discharge decreases with increasing distance, despite the fact that at least weak capacitive coupling persists. The decrease is attributed to the heterogeneous recombination of atoms to parent molecules on the surface of the borosilicate glass. The recombination efficiency is rather low for O and N atoms and extensive for H atoms. The dissociation fraction inside the intensive plasma in the H-mode is close to 100% for oxygen and hydrogen in the range of pressures up to several 10 Pa, whereas for nitrogen it is below 10%. Gas flow along the borosilicate glass tube affects the density of atoms away from the main discharge. As a general rule, a higher flow will result in better transfer of atoms from the source (i.e., powerful plasma) toward the remote parts of the discharge tube. The effect is particularly severe for hydrogen, where the decay length could be on the order of centimeters. Atom density on the order of 10^{21} m^{-3} is achievable in the E-mode (predominant capacitive coupling) for oxygen, whereas for nitrogen and hydrogen, high atom densities are achievable only in the H-mode (inductive coupling).

References

1. K. Tichmann, U. von Toussaint, W. Jacob, Determination of the sticking coefficient of energetic hydrocarbon molecules by molecular dynamics, *J. Nucl. Mater.*, **420** (2012) 291–296.
2. B. deB. Darwent, Bond dissociation energies in simple molecules, U.S. National Bureau of Standards, USA, Washington (1970).
3. P. Vašina, V. Kudrle, A. Tálský, P. Botoš, M. Mrázková, M. Meško, Simultaneous measurement of N and O densities in plasma afterglow

by means of NO titration, *Plasma Sources Sci. Technol.*, **13** (2004) 668–674.

4. M. Cacciatore, M. Rutigliano, Dynamics of plasma–surface processes: E–R and L–H atom recombination reactions, *Plasma Sources Sci. Technol.*, **18** (2009) 023002.
5. S. Markelj, I. Čadež, Production of vibrationally excited hydrogen molecules by atom recombination on Cu and W materials, *J. Chem. Phys.*, **134** (2011) 124707.
6. A. Vesel, Heterogeneous surface recombination of neutral nitrogen atoms, *Mater. Technol.*, **46** (2012) 7–12.
7. A. Drenik, A. Vesel, A. Kreter, M. Mozetic, Recombination of hydrogen atoms on fine-grain graphite, *Appl. Surf. Sci.*, **257** (2011) 5820–5825.
8. M. Mozetic, U. Cvelbar, A. Vesel, A. Ricard, D. Babic, I. Poberaj, A diagnostic method for real-time measurements of the density of nitrogen atoms in the postglow of an Ar-N_2 discharge using a catalytic probe, *J. Appl. Phys.*, **97** (2005) 103308.
9. I. Sorli, R. Rocak, Determination of atomic oxygen density with a nickel catalytic probe, *J. Vac. Sci. Technol., A*, **18** (2000) 338–342.
10. U. Cvelbar, M. Mozetic, A. Ricard, Characterization of oxygen plasma with a fiber optic catalytic probe and determination of recombination coefficients, *IEEE Trans. Plasma Sci.*, **33** (2005) 834–837.
11. A. Drenik, The probability of heterogeneous recombination of hydrogen and oxygen atoms on the surfaces of fusion-relevant materials, PhD thesis, Jozef Stefan International Postgraduate School, Ljubljana, Slovenia (2009).
12. R. Zaplotnik, A. Vesel, M. Mozetic, Atomic oxygen and hydrogen loss coefficient on functionalized polyethylene terephthalate, polystyrene, and polytetrafluoroethylene polymers, *Plasma Processes Polym.*, **15** (2018) e1800021.
13. A. Vesel, M. Mozetic, M. Balat-Pichelin, Reduction of a thin chromium oxide film on Inconel surface upon treatment with hydrogen plasma, *Appl. Surf. Sci.*, **387** (2016) 1140–1146.
14. A. Vesel, M. Mozetic, M. Balat-Pichelin, Sequential oxidation and reduction of tungsten/tungsten oxide, *Thin Solid Films*, **591** (2015) 174–181.
15. M. Mozetic, A. Vesel, J. Kovac, R. Zaplotnik, M. Modic, M. Balat-Pichelin, Formation and reduction of thin oxide films on a stainless steel surface upon subsequent treatments with oxygen and hydrogen plasma, *Thin Solid Films*, **591** (2015) 186–193.

16. B. J. Wood, H. Wise, Kinetics of hydrogen atom recombination on surfaces, *J. Phys. Chem.*, **65** (1961) 1976–1983.
17. M. Mozetic, A. Vesel, S. D. Stoica, S. Vizireanu, G. Dinescu, R. Zaplotnik, Oxygen atom loss coefficient of carbon nanowalls, *Appl. Surf. Sci.*, **333** (2015) 207–213.
18. M. Balat-Pichelin, J. M. Badie, R. Berjoan, P. Boubert, Recombination coefficient of atomic oxygen on ceramic materials under earth re-entry conditions by optical emission spectroscopy, *Chem. Phys.*, **291** (2003) 181–194.
19. M. Balat-Pichelin, D. Hernandez, G. Olalde, B. Rivoire, J. F. Robert, Concentrated solar energy as a diagnostic tool to study materials under extreme conditions, *J. Sol. Energy Eng.*, **124** (2002) 215–222.
20. P. Cauquot, S. Cavadias, J. Amouroux, Heat transfer from oxygen atoms recombination on silicon carbide. Chemical evolution of the material surface, *High Temp. Mater. Processes*, **4** (2000) 365–378.
21. T. Marshall, Surface recombination of nitrogen atoms upon quartz, *J. Chem. Phys.*, **37** (1962) 2501–2502.
22. Y. C. Kim, M. Boudart, Recombination of oxygen, nitrogen, and hydrogen atoms on silica: kinetics and mechanism, *Langmuir*, **7** (1991) 2999–3005.
23. G. A. Melin, A. Melin, R. J. Madix, Energy accommodation during oxygen atom recombination on metal surfaces, *Trans. Faraday Soc.*, **67** (1971) 198.
24. M. Balat-Pichelin, L. Bedra, O. Gerasimova, P. Boubert, Recombination of atomic oxygen on alpha-Al_2O_3 at high temperature under air microwave-induced plasma, *Chem. Phys.*, **340** (2007) 217–226.
25. M. Mozetic, A. Vesel, A. Drenik, I. Poberaj, D. Babic, Catalytic probes for measuring H distribution in remote parts of hydrogen plasma reactors, *J. Nucl. Mater.*, **363–365** (2007) 1457–1460.
26. R. Zaplotnik, A. Vesel, M. Mozetic, Transition from E to H mode in inductively coupled oxygen plasma: hysteresis and the behaviour of oxygen atom density, *EPL - Europhys. Lett.*, **95** (2011) 55001.
27. N. Krstulovic, I. Labazan, S. Milosevic, U. Cvelbar, A. Vesel, M. Mozetic, Optical emission spectra of RF oxygen plasma, *Mater. Technol.*, **38** (2004) 51–54.

Chapter 5

Surface Modification by Fusion Plasmas

M. Rubel,[a] S. Brezinsek,[b] and A. Widdowson[c]

[a]*KTH Royal Institute of Technology, Department of Fusion Plasma Physics, 10044 Stockholm, Sweden*

[b]*Forschungszentrum Jülich, Institute of Energy and Climate Research – Plasma Physics, D-52425 Jülich, Germany*

[c]*Culham Centre for Fusion Energy, Culham Science Centre, Abingdon, OX14 3DB, United Kingdom*

marek.rubel@ee.kth.se; s.brezinsek@fz-juelich.de; anna.widdowson@ukaea.uk

The focus of the chapter is on the consequences of plasma–material interactions (PMIs) in devices of controlled thermonuclear fusion. Particular emphasis is on processes involved in the interactions and their impact in terms of significant morphological changes in materials exposed to hot plasmas. The presentation is concise but with a holistic approach to the topic. Controlled fusion is briefly described, and nuclear reactions relevant from the point of view of reactor operation are introduced. This is followed by a presentation of reactor concepts and a broad description of plasma-facing components in fusion devices. Plasma-induced erosion processes are explained, and experimental tools and material analysis methods are addressed.

Plasma Applications for Material Modification: From Microelectronics to Biological Materials
Edited by Francisco L. Tabarés

ISBN 978-981-4877-35-0 (Hardcover), 978-1-003-11920-3 (eBook)
www.jennystanford.com

5.1 Introduction

The term "fusion" is derived from the Latin verb "fundare," meaning "melt" or "pour." In the middle of the sixteenth century, the Latin noun "fusio" became the French "fusion." According to the Oxford English Dictionary, it is "a process of joining two or more things together to form a single." In contemporary language, fusion has many meanings in various areas, from politics (e.g., union and coalition) to art (e.g., opera, jazz, and movies) and psychology (cognitive fusion of thoughts and reality) to culinary products (e.g., blending of ingredients of different regional cuisines). In the context of nuclear physics, the term "fusion" was first used in 1947. This chapter is devoted to this area, in particular to controlled thermonuclear fusion, which—in itself—is a broad interdisciplinary field where science and technology are integrated toward the main goal: construction and operation of a power-generating system to include fusion as a significant energy source in the future energy mix. This encompasses plasma and nuclear physics, materials science and engineering, chemistry, radiology, vacuum technology, signal processing, metrology, and many, many other branches.

The focus of this chapter is on plasma–material interactions (PMIs), in particular on processes involved in the interactions and the significant morphological changes arising from a material's exposure to hot plasmas. The presentation is concise but with a holistic approach to the topic. In the following sections, controlled thermonuclear fusion is briefly described and nuclear reactions relevant from the point of view of reactor operation are introduced. This is followed by a presentation of reactor concepts and broad description of plasma-facing components (PFCs) in fusion devices. Plasma-induced erosion processes and their consequences are described. The chapter concludes with a discussion on experimental tools and material analysis methods.

5.2 Controlled Thermonuclear Fusion: Reactions and Devices

There are at least two fundamental criteria for selecting a given nuclear process from the point of view of usefulness in energy production. The process has to be exothermic, and it should involve rather light nuclei to avoid very strong electrostatic repulsion between the reactants. As a result, reactions under consideration are those with deuterium (notation: D, d, ^{2}H), tritium (notation: T, t, ^{3}H), and helium-3 (notation: ^{3}He):

$$\mathrm{D + D \rightarrow T\ (1.01\ MeV) + H\ (3.03\ MeV)} \tag{5.1a}$$

$$\mathrm{D + D \rightarrow {}^{3}He\ (0.82\ MeV) + n\ (2.45\ MeV)} \tag{5.1b}$$

$$\mathrm{D + T \rightarrow} \alpha\ \mathrm{(3.52\ MeV) + n\ (14.06\ MeV)} \tag{5.2}$$

$$\mathrm{D + {}^{3}He \rightarrow} \alpha\ \mathrm{(3.67\ MeV) + H\ (14.69\ MeV)} \tag{5.3}$$

$$\mathrm{T + T \rightarrow} \alpha\ \mathrm{(3.77\ MeV) + 2n\ (7.53\ MeV)} \tag{5.4}$$

$$\mathrm{T + T \rightarrow} \alpha\ \mathrm{(4.03\ MeV) + H\ (4.03\ MeV) + n\ (4.03\ MeV)} \tag{5.5}$$

$$\mathrm{T + T \rightarrow} \alpha\ \mathrm{(4.76\ MeV) + 2H\ (9.54\ MeV)} \tag{5.6}$$

The branching ratio of reactions 5.1a and 5.1b is around 1. There are also reactions with lithium, especially D + ^{6}Li, but they are not discussed in the following.

Cross sections (σ) and energy release (Q value) of respective nuclear reactions indicate that the deuterium-tritium fusion, reaction 5.2, is the best possible option [1] because the cross section is greater than those of other processes listed above. The maximum is around 70 keV (700,000,000 K) of deuteron energy and a sufficiently high D-T reaction rate is reached already at 20 keV. The cross sections of other processes are distinctly smaller, and their maxima are at hundreds of kiloelectron volts, which would require efficient particle heating and their confinement at 2 billion kelvin or even higher. Therefore, at present, all plans and efforts toward commercial fusion are concentrated on a reactor fueled with heavy hydrogen isotopes deuterium and tritium.

From the Q value (17.58 MeV) and kinematics of the D-T reaction one concludes that burning 1 g of the equimolar D-T mixture results in 1.2×10^{23} fusion events releasing energy of 67.9 GJ and 271.8 GJ, associated respectively with α particles (helium-4 ions) and neutrons. Energy transfer from hot α particles to plasma is a prerequisite for efficient heating of reactants to achieve self-sustaining fusion. At the same time, the energy balance implies that the power transferred from α particles to the fuel must eventually be extracted by the wall—plasma-facing materials (PFMs) and

PFCs—that is, multimaterial structures, thus placing huge demands on the required material properties. Assuming that the operation of a system will release 1000 MW of fusion power, the reactor wall structures have to cope with about 800 MW (80%) associated with neutrons and 200 MW with α particles. The latter number is actually higher by around 50 MW when power injected by auxiliary heating systems, such as neutral beams and resonance methods, is taken into account. As a result, in total, 250 MW radiated by the plasma must be removed by PFCs. The energy of fast neutrons must be deposited in an absorber to facilitate (i) heat exchange and transfer to electricity generating systems of a power plant and (ii) tritium production, as described in Section 5.3. All this shows that issues related to primary (tritium) and induced radioactivity (neutron activation) and to power handling are universal for all types of systems realized for the generation and use of fusion energy.

The basic idea of a reactor is to compress fuel species and then to harvest the released energy by heat absorbers and heat exchange systems in the reactor wall. Two main schemes have been developed and tested for overcoming the Coulomb barrier and forcing nuclei to react: (i) magnetic confinement fusion of plasma ignited and maintained by a strong magnetic field of several tesla [2] and (ii) inertial confinement by intense laser [3] or ion beams [4]. Many scientific and technological issues are similar in all these approaches, in particular fuel production. In the following sections only the first scheme will be discussed, because it is the most mature from the reactor technology point of view.

There are two main concepts of controlled fusion devices (CFDs) acting as magnetic traps: tokamak and stellarator, both designed and constructed for the first time in the early 1950s. The tokamak was invented by Russian scientists Andrei Sakharov and Igor Tamm, and the device name is an acronym from that language meaning "toroidal chamber (camera) with magnetic coils." The concept is based on a transformer with a central solenoid where the plasma current in the toroidal chamber constitutes the secondary circuit [5]. The stellarator was invented by an American researcher Lyman Spitzer [6]. The device name refers to the possibility of harnessing reactions that power stars. Magnetic confinement is achieved in a stellarator with a strong magnetic field using only external coils encircling a toroidal vessel. The configuration is characterized by a "rotational transform" such that a single magnetic field line intersects a cross-sectional plane at points that successively rotate around the magnetic axis.

At present there are many (around 60) experimental fusion devices operated worldwide. The largest is the Joint European Torus (JET), at the Culham Science Centre (UK). Other big tokamaks are ASDEX[a] Upgrade (AUG), in Germany; KSTAR,[b] in the Republic of Korea; EAST,[c] in China; WEST,[d] in France; DIII-D, in the US; and JT-60SA, under construction until 2022, in Japan. There are also two large stellarator-type devices: LHD,[e] in Japan, and Wendelstein 7-X (W7-X), in Greifswald, Germany. The scientific and technological missions are defined for each machine: plasma performance, high energy confinement, optimization of plasma heating schemes, scenario development, test of actively cooled components, testing super-conducting (SC) operation, and wall materials.

5.3 Fusion Fuel and Reactor Components

Whichever energy generating system is used, the appropriate fuel is required. In the case of fusion these are gases for plasma production. Fundamental plasma studies can be performed with protium (^{1}H, hydrogen-1), as it does not lead to the generation of radioactive products. Therefore, hydrogen fuel is used during the initial prenuclear test phase of large machines in order to commission and check the functioning of operational systems, while advanced fusion studies require D and T. Current experimental CFDs are operated with deuterium (reactions 5.1a and 5.1b), but equimolar 1:1 D-T fuel was demonstrated in the 1990s at the JET [7–9] and Tokamak Fusion Test Reactor [10, 11].

Deuterium is not a radioactive isotope and has a natural abundance of 0.0115%. It can be extracted from water (~33.3 g D/m^3), meaning that seawater can provide deuterium in inexhaustible quantities. Tritium, however, is radioactive. It is a low-energy β^- emitter (maximum energy 18.59 keV, mean energy 5.7 keV), decaying

[a]Axially Symmetric Divertor Experiment (Garching, Germany)
[b]Korea Superconducting Tokamak Advanced Research (Deajoen, Republic of Korea)
[c]Experimental Advanced Superconducting Tokamak (Hefei, China)
[d]Tungsten (W) Environment in Steady-State Tokamak (Cadarache, France)
[e]Large Helical Device (Toki, Japan)

to ^{3}He with a half-life ($t_{1/2}$) of 12.323 years. The resulting specific radioactivity of 1 g T is equal to 9652 Ci (3.571×10^{14} Bq). In nature, tritium is formed as a product of the interaction of cosmic rays with atmospheric gases. Its total content in the environment, mainly the Earth's atmosphere, amounts to approximately 7.2 kg, while over 165 kg of T would be required per year for a reactor of 3 GW_{th} [12]. Therefore, tritium has to be produced, but options are scarce. Employing CANDU[f] reactors is not viable as the production rates are too low to meet the demands [13]. In addition, there are safety concerns regarding transportation of radioactive substances in large quantities. The solution is on-site tritium breeding in a lithium-containing mantel incorporated into the reactor design. Lithium has two stable isotopes, ^{6}Li and ^{7}Li, with a natural abundance of 7.59% and 92.41%, respectively. They are transmuted to tritium in three reactions:

$$^{6}\text{Li} + \text{n} \rightarrow \alpha + \text{T} + 4.78\ \text{MeV} \tag{5.7}$$

$$^{7}\text{Li} + \text{n} \rightarrow \alpha + \text{T} + \text{n}' - 2.47\ \text{MeV} \tag{5.8}$$

$$^{7}\text{Li} + \text{n} \rightarrow 2\text{T} + 2\text{n}' - 10.3\ \text{MeV} \tag{5.9}$$

Reaction 5.9 leads to the multiplication of neutrons but has a low cross section and only occurs under irradiation with high-energy neutrons. Thus it is clear that the breeder must be heavily enriched with ^{6}Li. A variety of compounds have been tested, such as lithium ceramics (e.g., oxides, aluminates, siliconates, titanates, and zirconates), a mix of Li and Be fluorides (so-called Flibe), and also lithium-lead eutectic alloys in solid or liquid form. Both Flibe and Li-Pb contain neutron multipliers (Be and Pb, respectively), necessary to enhance the T breeding ratio [12, 14]. Several concepts of tritium breeding modules (TBM) with water or helium cooling have been proposed and are under development (e.g., Refs. [15, 16]), but the ultimate test of their efficiency and performance will only be possible in a reactor-class device operated with D-T providing sufficient neutron fluxes. The next-step fusion machine of a reactor class is the International Thermonuclear Experimental Reactor (ITER), currently under construction in Cadarache, France. The objectives of the ITER science and technology program are (i) an extended burn time of about 400 s, (ii) achievement of a self-sustained thermonuclear burn, (iii) safe operation of a reactor-like

[f]CANada Deuterium Uranium

device, (iv) testing of components under reactor-like conditions, and (v) testing of tritium breeding blanket modules [17, 18]. Six TBM concepts are planned to be tested on the ITER over a period of several years of D-T operation.

In brief, a reactor will be composed of a support structure, a cryostat, and SC magnets surrounding the vacuum vessel, with the first wall being an integrated blanket. The integrated blanket includes structural materials, neutron absorber and neutron multiplier materials, and high-heat flux components (i.e., plasma-facing armor brazed to heat sink plates or pipes). This is accompanied by tens of essential auxiliary systems providing fuel and other gases, vacuum conditions, cooling agents (water, gases also in liquid form), and plasma heating.

5.4 Plasma-Facing Wall

A collection of images in Figs. 5.1–5.6 shows the complexity of in-vessel structures in two tokamaks, TEXTOR[g] and JET, and in the W7-X stellarator. The selection of TEXTOR and JET is not accidental. These tokamaks belong to two distinct categories from the point of view of geometry, scientific and technological missions, and control of PMI processes; the processes and their consequences are described in Section 5.5. For this chapter, an important difference between the two tokamaks is related to plasma edge control using the so-called limiters and divertor [2]. Their role in plasma edge control results in strong interaction with the plasma. Therefore, the surfaces of these components are under the highest heat and particle loads. The limiter is defined as a shaped material block protruding from the main wall and used to intercept particles escaping the plasma. It separates the plasma from the scrape-off layer (SOL), that is, the region between the vessel wall and the last closed flux surface (LCFS), the position of which corresponds to the innermost part (tip) of the limiter and by which, the plasma radius (a) is defined. The divertor is a separate region in the vacuum vessel to which escaping particles are exhausted by means of extra magnetic field (extra coils). Edge control with limiters only calls for a circular cross section of a vacuum vessel. Divertor machines are D-shaped, but

[g]Toroidal Experiment for Technology Oriented Research (Jülich, Germany, until 2013)

limiters are implemented in the main chambers of such devices. They are located in the main part of the chamber.

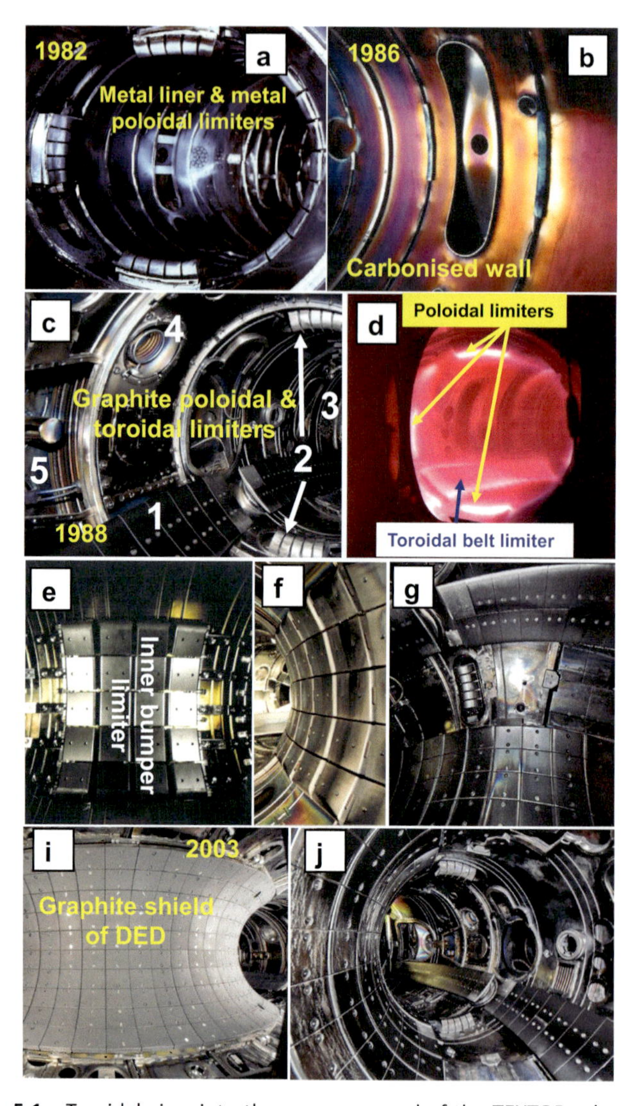

Figure 5.1 Toroidal view into the vacuum vessel of the TEXTOR tokamak: an overview of the development of plasma-facing material components over the year. The view during plasma discharge is in frame (d). In frame (e), numbers 1 and 2 indicate, respectively, the toroidal belt and poloidal limiters; 3 is the antenna for ion cyclotron radio frequency heating (ICRH); 4 the antenna for glow discharge operation during wall conditioning; and 5 indicates the liner.

TEXTOR, which operated for 30 years, from 1982 to 2013, at the Forschungszentrum Jülich, was a medium-sized machine of a circular poloidal cross section (minor radius a = 0.5 m) and with arrays of limiters [19]. At the beginning, it was a full metal device with the Inconel plasma-facing wall (called the liner; vessel-in-vessel) and three arrays of poloidal limiters made of stainless steel, as presented in Fig. 5.1a. Plasma contamination by metals eroded from the wall led to the search for low-Z materials and, as a result, large-scale wall conditioning techniques were developed for the first time to coat the inner surface area of the TEXTOR vacuum chamber. Coating with low-Z protective films started in year 1985, with carbonization [20], which was followed by boronization [21] and siliconization [22]. A colorful pattern arising from the thickness, 150–400 nm, of fringes in the carbonized layer is shown in Fig. 5.1b. Ultimately, metal limiters were replaced by carbon (graphite and carbon fiber composites [CFCs]) to achieve low-Z plasma-facing surfaces and an actively pumped toroidal belt limiter was installed: Advanced Limiter Test II (ALT-II), the main PFC in TEXTOR, shown in Fig. 5.1c. The image in Fig. 5.1d was taken during a tokamaks discharge to illustrate plasma interaction with limiter surfaces. A description of the in-vessel structures is provided in Fig. 5.1c. Numbers 1 and 2 indicate respective limiters: a toroidal belt and two poloidal arrays composed of five graphite blocks each. Number 3 denotes the antenna of radio frequency heating (ICRH), while 4 is the antenna for glow discharge operation enabling wall conditioning; 5 indicates the liner. In 1994 the wall protection on the high field side was added (Figs. 5.1e–5.1g): a bumper limiter composed of 10 × 10 cm graphite blocks. The image in Fig. 5.1g was taken from top of the machine, thus showing the lower tiles of the bumper limiter, the bottom poloidal limiter, and the liner. All consecutive upgrades and wall conditioning techniques were implemented to substantially increase the operation window and pulse duration. The last major change to the TEXTOR wall structure was to install additional coils on the high field side, the so-called dynamic ergodic divertor. This required robust protection of coils and corresponding significant extension of the inner bumper limiter covering about 30% of the plasma-facing wall. It is shown in Figs. 5.1i and 5.1j in the initial stage and after a few years of operation, respectively. Details regarding the internal components of TEXTOR are also presented in Section 5.6.

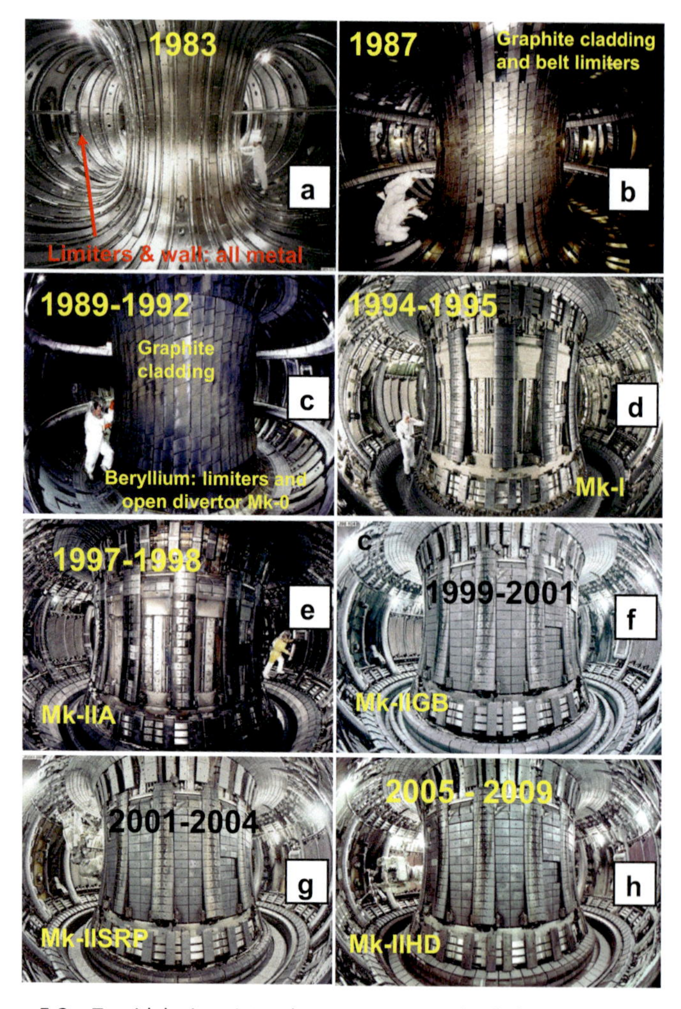

Figure 5.2 Toroidal view into the vacuum vessel of the JET tokamak: an overview of the development of plasma-facing material components over the year. Inconel wall and limiters [23] (a); graphite cladding and limiters [23] (b); beryllium limiters and tiles of the open divertor Mk-0 (c); graphite limiters and Mk-I divertor operated first with graphite and then with beryllium divertor tiles (d); graphite limiters and series of Mk-II divertors with graphite tiles: Mk-IIA (e); Mk-II Gas Box [72] (f); Mk-II Septum Replacement Plate [23] (g); Mk-II High Delta [77] (h).

The main mission of JET is the development of high-confinement regimes for next-step devices, that is, ITER. This also includes testing of divertor structures. Therefore, JET is a D-shaped tokamak. However, in June 1983, it started operation as a limiter machine with an Inconel plasma-facing wall and relatively small limiters, Fig. 5.2a. A few years later, the wall was covered by carbon tiles and two toroidal belt limiters composed of over 2000 CFC blocks were installed—see Fig. 5.2b. In 1989 the use of beryllium on a large scale was introduced: wall conditioning by Be evaporation and bulk metal belt limiters and floor tiles (Fig. 5.2c). Beryllium-castellated belt limiter tiles are shown in Fig. 5.3a. Castellation, that is, narrow (0.4–0.6 mm) grooves, is thought to be the best solution to ensure the thermomechanical durability and integrity of materials under high heat flux loads, especially when considering the use of metals. With the installation of the divertor coils and powerful ICRH antennae, the wall structure was distinctly changed. In Fig. 5.2d one perceives a channel at the bottom of the vessel—a Mark-I (Mk-I) divertor—operated first with CFC and then with castellated Be tiles. The appearance of Be tiles is shown in Fig. 5.3b; the CFC tiles of that divertor were of the same shape and size, but without the castellation. Further changes that can be visualized in this photographic survey have been mainly associated with the divertor structure and with in-vessel contamination by tritium. In consecutive divertors (Fig. 5.2e–g), the tiles were made of CFC to ensure efficient power handling. Figure 5.2e, taken in 1997, shows the in-vessel work carried out manually. After a full deuterium-tritium campaign in 1997 (May – November) all in-vessel operations have been performed by remotely handled robots; they are visible on the left side in Figs. 5.2g and 5.2h. Differences between consecutive divertors are to a large extent connected with the arrangement of the bottom tiles: a broad A-shape in Mk-IIA, a W-shape in Mk-II Gas Box (with the septum, that is, the divider plate between inner and outer divertors), a flat tile in Mk-II Septum Replacement Plate, and an inclined configuration in Mk-II High Delta. A detailed description of the shape of divertors in Fig. 5.2e–h and corresponding plasma footprints have been presented in a dedicated overview by Coad et al. [23]. CFC tiles from the divertor floor and a Gas Box structural element are shown in Figs. 5.3c and 5.3d, respectively.

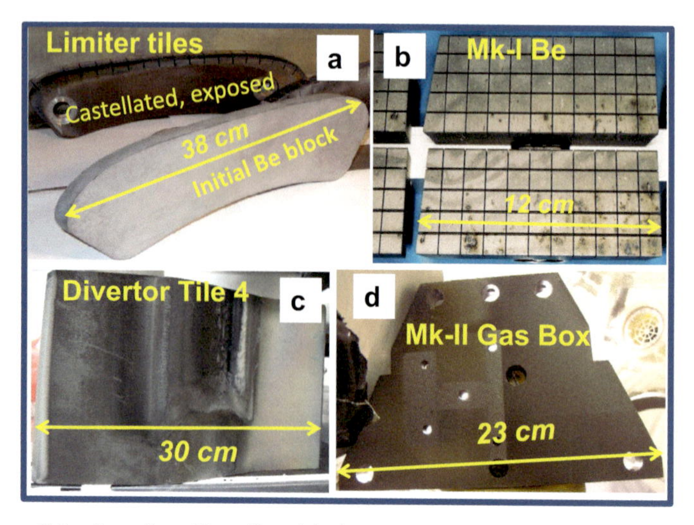

Figure 5.3 Castellated beryllium blocks used in the Mk-0 operation (a) and in the water-cooled Mk-I Be divertor (b); a divertor tile from the Mk-IISRP structure (c) and the element of the Gas Box from the Mk-IIGB structure (d).

Figure 5.4 Toroidal view into the vacuum vessel of the JET tokamak with the ITER-Like Wall: beryllium on the main chamber wall and tungsten in the divertor.

The long-term operation of JET with a carbon wall (JET-C) led to the development of high-confinement scenarios but also indicated negative aspects of the carbon surrounding: accumulation of large quantities of fuel in PFC and in remote areas of the divertor, that is, zones shadowed from the direct plasma line of sight [23]. This would result in unacceptably high levels of tritium inventory in a reactor-class machine fueled by a D-T mixture. Consequently, the decision was taken to completely refurbish the plasma-facing wall in JET. The carbon wall components were removed and replaced by beryllium in the main chamber and tungsten in the divertor [24]. Details regarding this transition are in Section 5.6. The toroidal view of the inside of the vacuum vessel of JET with Be and W components is shown in Fig. 5.4. Representative images of wall components are collected in Fig. 5.5. The castellated bulk beryllium inner wall guard limiter and the upper dump plate are in Figs. 5.5a and 5.5b. Castellation of the surface, that is, 12 mm deep and 0.4 mm wide grooves in the bulk material, is done to ensure thermomechanical durability of wall tiles under high head loads and to reduce halo currents; all PFCs in ITER will be manufactured in this way. Figures 5.5c and 5.5d show typical divertor components, such as a tungsten-coated (20 μm W layer) CFC divertor tile and a divertor module with bulk tungsten lamellae.

W7-X is an advanced stellarator operated since 2016 at the Max-Planck Institute for Plasma Physics in Greifswald, Germany [25, 26]. It is a toroidal modular system with five segments of nonuniform poloidal cross sections. The stainless-steel wall is protected by a complex structure of graphite and CFC tiles, as shown by an overview image in Fig. 5.6a, while images in Figs. 5.6b and 5.6c demonstrate the difference in wall appearance before and after plasma operation.

In summary, the collection of images shown here has taken the reader on a "historical tour" of over 35 years of developments in the wall technology of magnetic CFDs. The images document the many crucial structural changes that have been implemented over the years. From this fact, one may infer the paramount significance of PFMs and components for reactor operation. Most changes have been introduced to improve control over plasma–wall interactions (PWIs)—one of the primary areas where integration of physics and technology is being achieved.

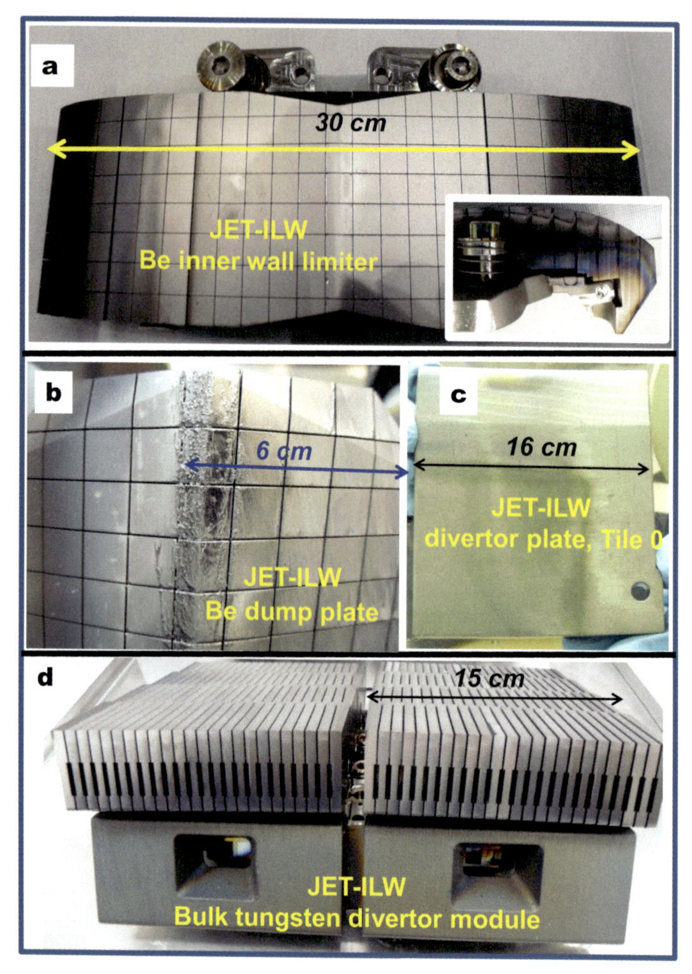

Figure 5.5 JET-ILW. Castellated beryllium limiters from the high-field side, that is, the inner wall; the insert shows deposition on the side surface (a) and the upper dump plate with molten areas (b), the W-coated CFC plate from the inner divertor with a deposition pattern (c), and the bulk tungsten lamellae module from the divertor base (d).

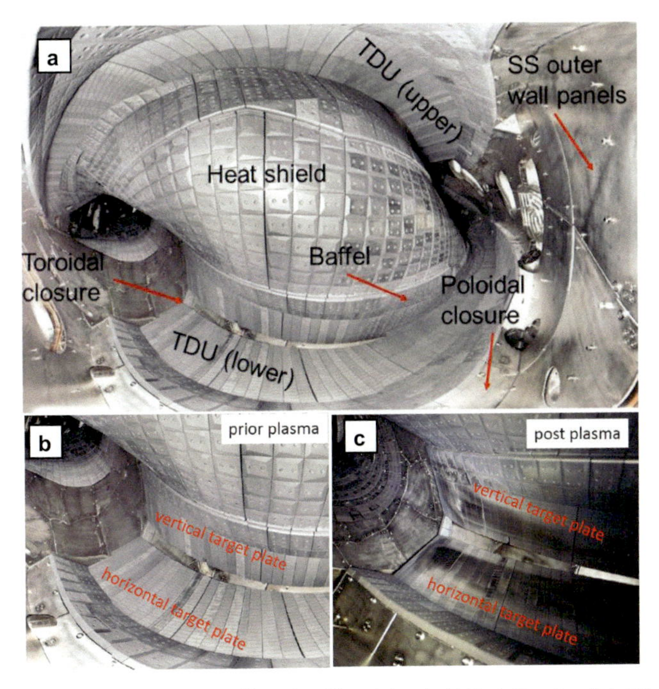

Figure 5.6 Plasma-facing wall in the Wendelstein 7-X stellarator (a). Divertor target plates before (b) and after plasma operation (c).

5.5 Plasma–Wall Interactions

The aim of this section is to introduce a spectrum of phenomena responsible for and related to the modification of in-vessel materials and components. PWIs comprise all processes involved in the exchange of mass and energy between the plasma and the surrounding wall. A description of the resultant effects requires proper in situ diagnosis of in-vessel processes by a range of spectroscopic and thermal techniques in conjunction with ex situ analytical tools to determine changes in the morphology of the plasma-exposed materials.

Energy leaves plasma in the form of electromagnetic radiation and kinetic energy of particles. A brief overview of the processes is schematically shown in Fig. 5.7. The plasma-surrounding wall is irradiated with ions in different charge states, charge-exchange neutrals, electrons, neutrons, and photons, with a very broad spectral range, from one-tenth of a nanometer to tens of micrometers. They originate from nuclear (γ) and electronic processes (X-ray, UV, visible). All incident particles modify material properties, ranging from the surface to the bulk. While properties of materials are changed by particle fluxes hitting the wall, the plasma becomes contaminated by species eroded from the wall. Atoms removed are instantly ionized and move along the magnetic field lines. Eventually, if not pumped out of the vacuum vessel in the form of volatile molecules, eroded species migrate to the plasma edge and are re-deposited at a location other than the place of origin. Re-deposition of plasma impurity species is accompanied by co-deposition with fuel atoms, that is, hydrogen isotopes. As a result of material migration, for example, co-deposition and mixed layer formation, physical, chemical, and thermomechanical properties of PFCs are changed. The complexity and consequences of PWI processes have been reviewed in Refs. [27–29].

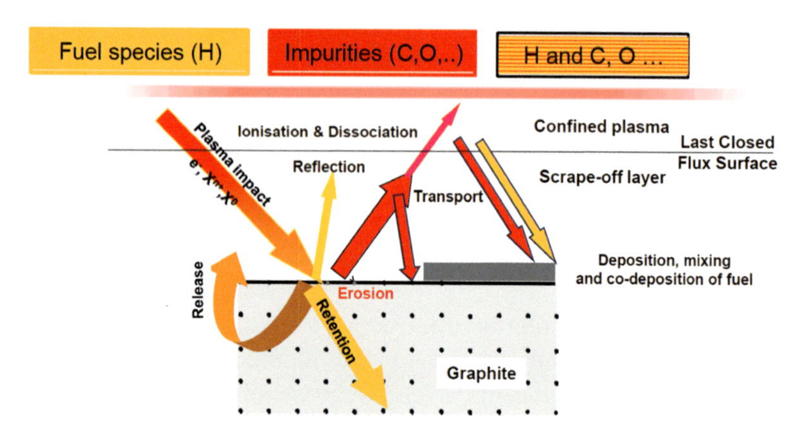

Figure 5.7 Schematic illustration of plasma-wall interactions processes. An example of the carbon wall is shown.

Two interrelated aspects of fusion reactor operation—economy and safety—are the driving forces for PWI studies and define the major objectives of research in this area:

- Lifetime of materials and components
- Detailed assessment of accumulation of hydrogen isotopes in PFCs, especially tritium inventory, because it is crucial for licensing (safety)

- Dust formation as a result of material erosion, especially in the case of tritiated and neutron-activated particles
- Plasma impact on diagnostic components that are decisive for plasma diagnosis and are crucial in machine protection systems

Therefore, the challenges are at least twofold: (i) to determine plasma composition as it has a direct impact on reactor performance; (ii) to examine in detail the state of wall components (surface and bulk) as it is the only way to understand and tackle major objectives of PWI studies.

5.6 Erosion Processes and Wall Materials

Erosion is defined as a change in material morphology—the surface, the bulk, or both—under the impact of an ambient environment. The spectrum is broad, from the modification of the outermost surface layer structure and composition to material losses and dramatic degradation of properties. Nature teaches us that there is no erosion-resistant material. It is only a matter of material properties and exposure conditions. This is also the case in fusion-related materials science, that is, PWI studies. Erosion processes and their consequences for plasma are fully connected with the materials used in wall technology. Therefore, in this section, the required properties of materials are introduced and the various erosion processes described.

5.6.1 Selection of Plasma-Facing Materials

The search for suitable wall materials had already started in the late 1960s when the detrimental effects of PWI on plasma performance were recognized. When saying "detrimental," one has to stress that PWI processes are, firstly, unavoidable because the plasma is surrounded by the wall of a vacuum vessel and, secondly, necessary for helium thermalization and for the exhaust and extraction of neutron energy in the reactor blanket. Therefore, materials of the integrated blanket (i.e., PFC and neutron absorber) must be compatible with ultrahigh vacuum, coolant fluids in the respective heat sinks, cryogenics (cryopumps), magnetohydrodynamics, neutron irradiation, thermomechanical stability, and durability.

A list of properties required for materials directly facing plasmas comprises many additional points: high thermal conductivity for handling high heat loads; resilience to thermal shocks; high melting point; low erosion yield under a range of particle energies; low-Z to reduce energy radiation losses by plasma contaminants; low affinity with hydrogen and oxygen toward chemical erosion, resulting in the formation of volatile products; low sorption of hydrogen isotopes; and high affinity to plasma impurity species (e.g., oxygen) toward the formation of stable nonvolatile compounds leading to oxygen gettering. Another indispensable feature is the compatibility of PFM with the heat sink structure (mainly Cu-Cr-Zr alloys) to ensure robust and reliable active cooling. There is no single element, compound, alloy, or composite satisfying all criteria. The physics underlying erosion cannot be bypassed, hence the efforts in plasma edge engineering are focused at the improved control of interactions by optimizing reactor operation scenarios and selection of the best-possible and available wall materials.

Over the years, a variety of materials have been tested, but only a few of them have been used for wall components in CFDs. The status until the end of the twentieth century is very well summarized in Table 1 of Ref. [28]. The main interest has been on carbon (C) in the form of graphite or various CFCs, tungsten (W), and beryllium; therefore processes described in the next paragraph will focus mainly on these elements.

As shown in Figs. 5.1, 5.2, and 5.4, for many years carbon has been the main wall material in most devices. Its power-handling capabilities are excellent, but its affinity to hydrogen isotopes results in chemical erosion (formation of hydrocarbons) and, as a consequence, re-deposition and build-up of co-deposited layers with a high content of fuel species, leading to unacceptably high fuel inventory (see Section 5.6.2). To address these issues, a large-scale test with an all-metal wall was decided in year 2004 at the JET tokamak: ITER-Like Wall (JET-ILW). Carbon components (JET-C operated till October 2009, as indicated in Fig. 5.2) were replaced by beryllium on the main chamber wall and tungsten in the divertor [24, 30]—see Figs. 5.4 and 5.5. The project to implement JET-ILW involved a very broad R&D program and was completed in May 2011. A few months later, the tokamak operation began. Elimination of the carbon source indeed led to a significant decrease in fuel retention [30–32]. As a consequence, a decision was made by the ITER Organization to abandon carbon PFCs and operate the reactor only with metallic walls: Be and W [33].

5.6.2 Erosion and Deposition

Erosion processes in fusion devices can be categorized in several ways, depending either on the type of materials undergoing certain erosion pathway or on the energy involved in the process. In the second category one distinguishes phenomena under quiescent steady-state operation from those under high-power loads and especially off-normal events, such as disruptions and vertical displacements.

A wide spectrum of processes governing wall erosion and modification comprises:

- Physical sputtering, which occurs for all types of materials
- Chemical erosion (chemically enhanced sputtering) by fuel species, which occurs for carbon forming hydrocarbons
- Chemical erosion by oxygen impurities forming volatile oxides: CO_x, WO_3, etc.
- Sublimation and brittle destruction, which occur for carbon under high power loads
- Melting and splashing of the melt layer in the case of metals
- Arcing
- Neutron-induced modification: transmutation and volumetric damage with a whole set of radiation effects, which occur for all materials

5.6.2.1 Erosion-deposition under steady-state conditions

Physical sputtering is referred to as the removal of surface atoms by incoming energetic particles. The quantitative effect is determined by the sputtering yield, ${}^{\mathrm{T}}Y_{\mathrm{P}}$, defined as the number of atoms removed from the target per incident particle (projectile) either ionized or neutral, where T denotes the target atom and P the incident particle. The value depends on the masses in the system (M_{P} and M_{T}), the projectile energy, the angle of incidence, and the surface state, especially roughness. Target temperature is of secondary importance, unless it is close to the melting point. Sputtering and, therefore, erosion occurs for all types of materials. The maximum effect is measured at the incident angle of 30°–45° with respect to the surface in the energy range of 1–30 keV. Depending on the projectile-target combination, the values of $^{T}Y_{P}$ are between 5 × 10^{-3} (light projectile on a high-Z target) and around 10 (high-Z metallic impurity bombarding, for instance, a Be or C surface), thus indicating how important both plasma edge control and PFC shaping are in reducing erosion. For example, the threshold for Be sputtering by deuterium is on the order of a few electron volts, while tungsten is eroded by D species above 200 eV. The physics of sputtering and experimental aspects are comprehensively presented in Refs. [34, 35], while the consequences of sputter erosion for wall materials are in Refs. [27, 28, 36]. Sputtering yields can be calculated with advanced codes such as the Stopping and Ranges of Ion in Matter [37].

Chemical erosion, also called "chemically enhanced sputtering," takes place when the incident flux reacts chemically with target atoms, forming volatile products instantly leaving the exposed surface. Quantitative effects of erosion depend on particle energy and flux density, but also, like for all chemical reactions, target temperature and surface structure are crucial factors. The affinity of carbon for hydrogen isotopes ensures that carbon-based PFCs are affected by the flux toward the formation of various hydrocarbons (C_xH_y with a majority of C_1 and C_2 molecules), which can be transported locally or over long distances [29]. Re-deposition of eroded material leads to the formation of co-deposits containing also vast amounts of hydrogen isotopes, eventually decisive for radioactive tritium inventory in the case of D-T operation. Images and a graph in Fig. 5.8 demonstrate the problem. On the surface of the toroidal pump limiter ALT-II from TEXTOR (Fig. 5.8a), one perceives shiny and blackish areas, which are, respectively, erosion (E) and deposition (D) zones. On a single limiter tile (Fig. 5.6b,c) the surface in the erosion zone is smooth, while in the other area co-deposited layers tend to peel off, forming dust. The deuterium and boron (from boronizations) contents measured by nuclear reaction analysis (NRA) are not uniform (Fig. 5.8b), showing the difference between the D amounts in the erosion zone (5 × 10^{17} cm^{-2}) and the deposition area (50 × 10^{17} cm^{-2}). The actual difference in the fuel content is over 2 orders of magnitude, because the co-deposit thickness was nearly 50 μm, while only 5 μm could be measured by NRA [38]. The structure of flakes and co-deposits is very diverse, as shown in electron micrographs in Fig. 5.8d–g. However, the main point is the highly developed surface topography, which means a high surface area. In addition, the layers are porous and brittle, creating the risk for further disintegration into fine dust particles of micrometer size.

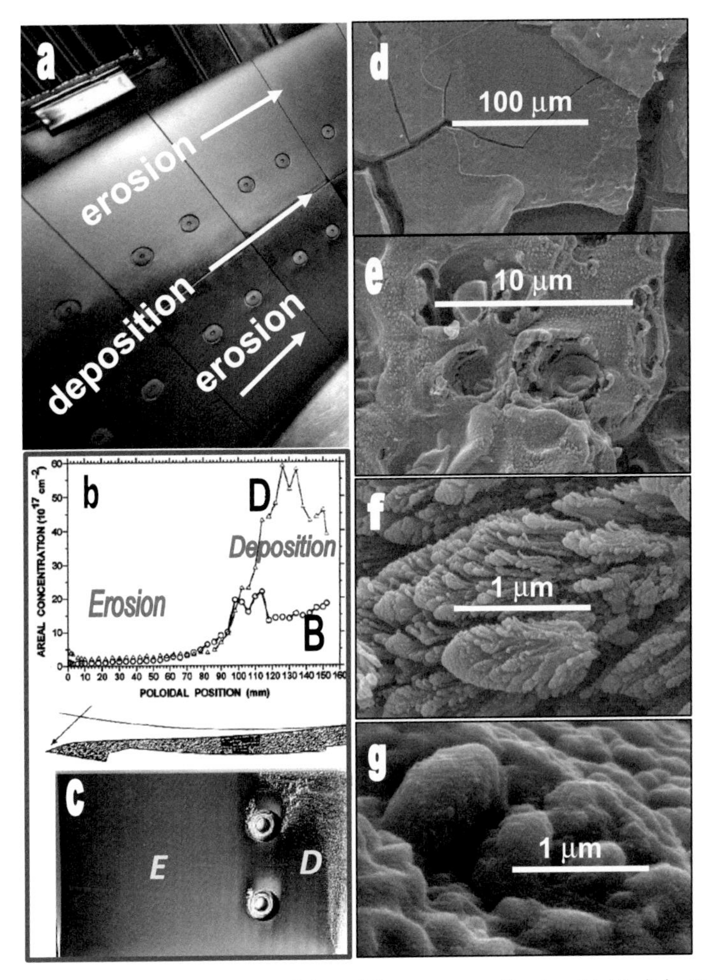

Figure 5.8 TEXTOR. Erosion and deposition zones on the toroidal belt limiter ALT-II (a), surface content and distribution of deuterium and boron on the limiter plate (b), the limiter plate with erosion (*E*) and deposition zones (*D*) where deuterium and boron were determined (c), and general features and detailed structures of deposits on the limiter plate (d–g).

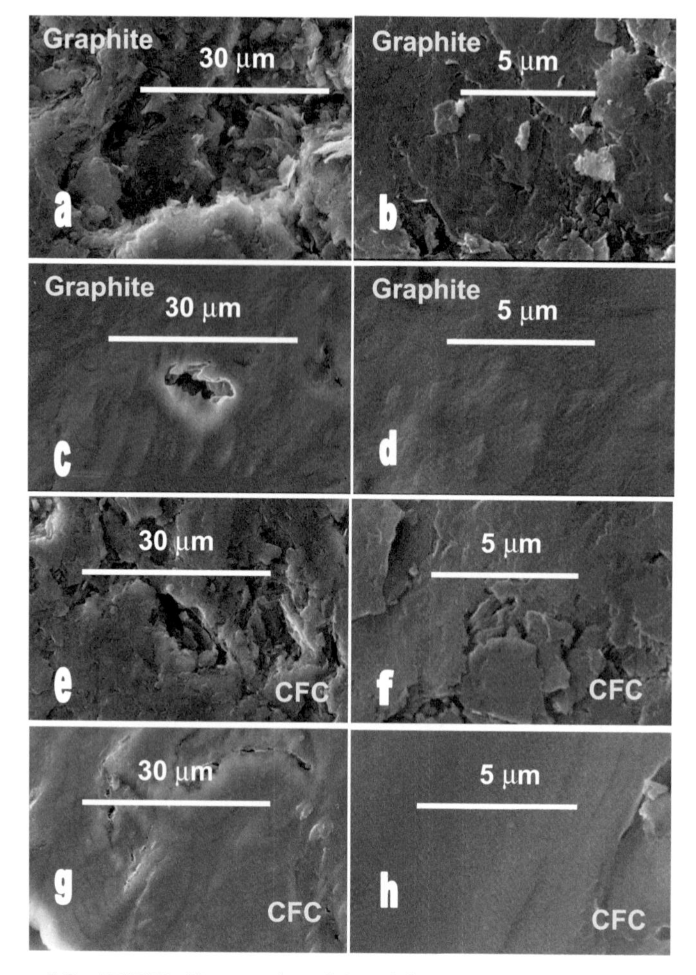

Figure 5.9 TEXTOR. Topography of initial (unexposed) and plasma-eroded graphite and CFC tiles on the toroidal belt limiter ALT-II. High-resolution SEM images of (a, b) unexposed graphite, (c, d) erosion zone on a graphite plate exposed to the plasma for 15,000 s, (e, f) unexposed CFC tile, and (g, h) erosion zone on a CFC plate exposed for 15,000 s.

On the contrary, erosion zones become smoothed by plasma impact, as documented in Fig. 5.9, for two types of ALT-II tiles: the central plates, made of graphite, and the corner blocks, of CFC, as marked by red dots in Fig. 5.8a. The frames in Figs. 5.9a,b and 5.9e,f show at two different magnifications the unexposed tiles of two types. The original surfaces are rough, while after exposure to

15,000 s of plasma operation, the surfaces show a distinct effect of smoothening, as shown in Figs. 5.9c,d and 5.9g,h for respective areas even on CFC tiles [39].

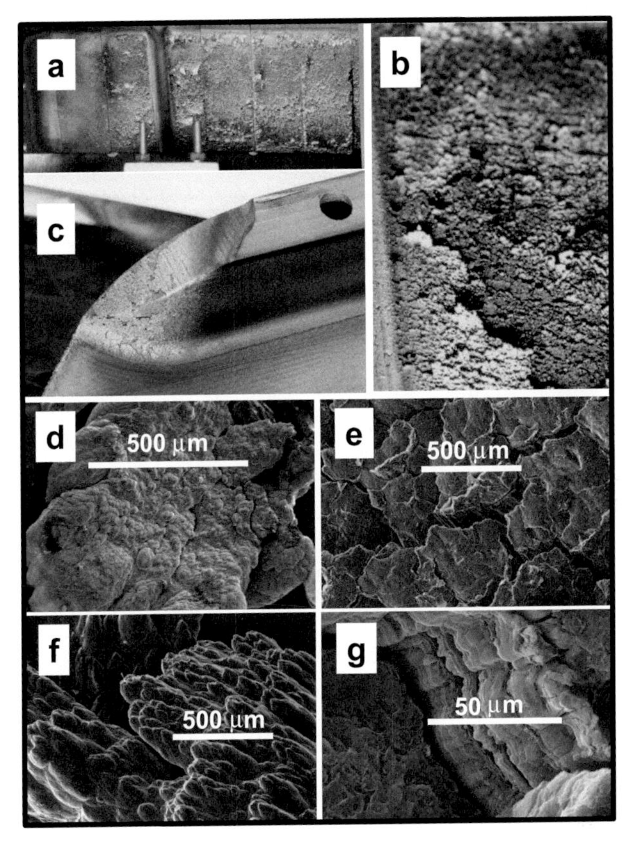

Figure 5.10 TEXTOR. Neutralizer plates of the toroidal belt limiter ALT-II after 95,000 s of operation (a); overview of 1 mm thick flaking co-deposits (b); edge of the skimmer (c) and detailed structures of co-deposits (d–g).

Transport and re-deposition are not limited to PFC surfaces. Gases are actively pumped out, but hydrocarbons are deposited on the way to the pump duct. Images in Fig. 5.10a and 5.10b show neutralizer plates and the skimmer (Fig. 5.10c) of the ALT-II pump limiter after around 90,000 s of operation. There is a brittle crust of carbon with some admixtures of boron from boronizations. Different co-deposit structures are in micrographs in Fig. 5.10d–g. The layers of the mixed microstructure reach 1 mm in thickness, the thickest ever recorded. Similar, though not so thick, co-deposits were also formed on protection limiters of the ICRF antenna shown in Fig. 5.11a. Significantly less deposition was found on the inner bumper limiter, 1×10^{18} cm^{-2} (Fig. 5.11b). The formation of carbon-rich deposits is also seen on all other components from carbon wall machines: tungsten-coated main poloidal limiters from TEXTOR (Fig. 5.11c) and the test poloidal limiter in Fig. 5.11e. A wall tile and a probe head from W7-X are, respectively, in Figs. 5.12a and 5.12b.

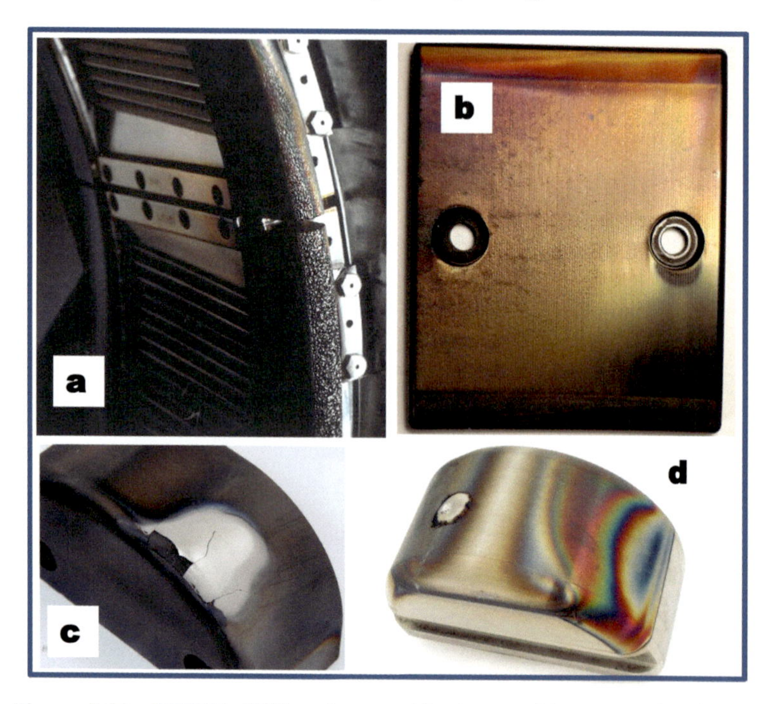

Figure 5.11 TEXTOR. ICRH antenna with the graphite protection limiter covered by thick deposits (a); a graphite tile of the inner bumper limiter (b), a graphite block of the main poloidal limiter coated with a vacuum plasma sprayed tungsten layer damaged during high power loading tests (c); test poloidal limiter with a deposition pattern (d).

Oxygen impurities in fusion plasmas (0.5%–1%) are responsible for the formation of CO and CO_2. Volatile oxides may also be formed with tungsten (WO_3), even in a hydrogen-rich atmosphere [40], if there is a sufficient local source of oxygen, as was shown in experiments with power loading of bulk tungsten limiters in

TEXTOR [41]. Migration of tungsten due to the volatility of oxides is a concern in the case of air or water ingress into the vacuum vessel during operation with a hot tungsten PFC [42].

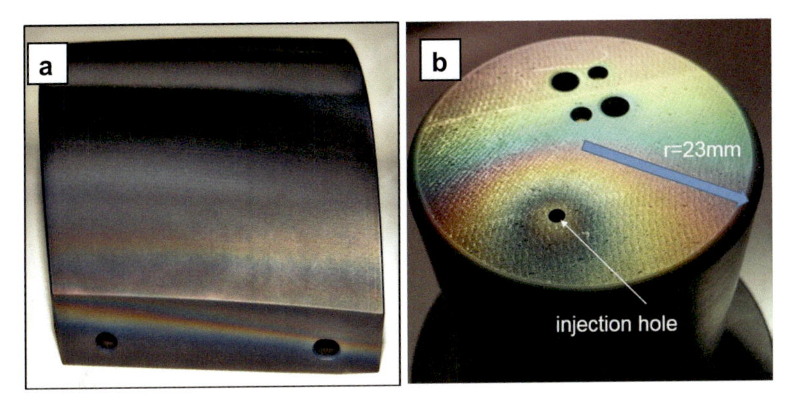

Figure 5.12 Wendelstein 7-X. a graphite tile after the first experimental campaign (a) and a probe head used during that campaign for electrical measurements (b).

Physical sputtering and chemically enhanced sputtering are the major mechanisms for wall erosion under quiescent steady-state operation. Physical effects can be predicted, but theoretical prediction of chemical erosion rates is very difficult, especially in the case of re-erosion of co-deposited layers whose properties are distinctly different from those of pure bulk materials. The field requires very strong experimental support; it will be briefly addressed in Sections 5.7 and 5.8. The data presented above have clearly shown the diversity of carbon erosion effects. High erosion rates and unacceptable levels of fuel accumulation in co-deposited layers have been the main drivers for the ITER-Like Wall in JET and for the ITER decision to eliminate carbon PFCs.

Erosion and deposition always occur simultaneously, and the mass balance must be maintained. As a result, all erosion phenomena are followed by material migration and deposition of the eroded material at a nearby location (local migration) or a distant location (long-distance migration). One discusses a net effect, which is defined as the difference between the gross erosion, the local re-deposition of the eroded material, and the deposition of impurities originating from other areas. This represents a difference of large numbers, describing the gross erosion and the gross material deposition. In the assessment one takes into account particle fluxes reaching the target, that is, both hydrogen isotopes (Γ_H) and respective impurities ($\Sigma\ \Gamma_I$). The latter flux is diminished due to the projectile reflection described quantitatively by reflection coefficients, R_I. With this, one assesses the amount of species deposited (e.g., implanted) in the target while with the sputtering yields, the amount of erosion products is calculated: ($\Gamma_H Y_H$) + ($\Sigma\,\Gamma_I Y_I$). Reflection coefficients and sputtering yields for all projectile-target combinations are dependent on the ion energy dependent on the edge electron temperature (T_e), which decides the sheath potential and, by this, the ion acceleration to the wall [2]. Finally, the ratio of erosion (E) to deposition (D) can be obtained:

$$E/D = (\Gamma_H Y_H) + (\Sigma\,\Gamma_I Y_I)/\Sigma\,\Gamma_I(1 - R_I) \qquad (5.10)$$

Results over 1 indicate net erosion, while values below 1 indicate net deposition. Even in a particular case of quantitative equilibrium (E/D = 1), the plasma-exposed surface is modified by hydrogen isotopes and impurities.

Prompt re-deposition is a particular case of local re-deposition. It occurs when the penetration depth of sputtered neutrals into the plasma becomes as small as the ion Larmor radius, or even smaller than that. Under such conditions, re-deposition is likely to take place within the first gyration of the ion [43]. Important parameters are the penetration depths of impurities, the local geometry of a target with respect to the magnetic field lines, and possible forces acting on the ionized impurities. High-Z impurities have a small penetration depth due to the low first ionization potential and low velocity related to their large mass. For instance, for W sputtered from a target under a T_e of 60 eV and an edge electron density, n_e, of 5×10^{18} m^{-3}, the mean free path, $\lambda_{ionization}$, for the ionization of W neutrals of 5–8 eV is around 0.2 cm and this length is equal to the Larmor radius under these conditions. It certainly reduces the net erosion of tungsten and its penetration into the plasma but does not prevent the effective migration of that species. Re-deposited atoms are re-eroded, and this cycle is repeated, leading to a "spillover" of tungsten until the equilibrium between the re-erosion and the re-deposition is reached. This is illustrated in Fig. 5.13 by results of a tracer (see Section 5.7.2 for information about tracers) experiment involving

W atoms deliberately injected into a series of tokamak discharges in TEXTOR [44]. A plasma flux containing both deuterium and all impurity species arrives at the net erosion zone of the limiter. The "wavy" profile of tungsten distribution on the ALT-II limiter plates determined by surface analyses indicates the gradual W transport toward the net deposition zone in a multistep sequence of primary deposition – primary erosion – prompt re-deposition – re-erosion – prompt re-deposition, etc. In summary, the control of erosion and deposition is crucial for a steady-state operation of future devices because it is decisive for material lifetime and fuel inventory.

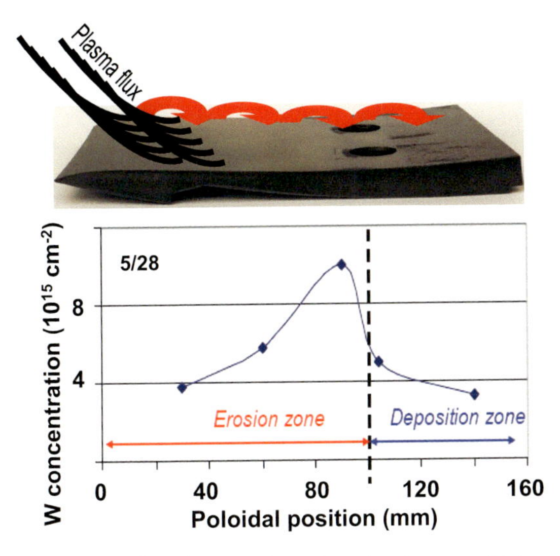

Figure 5.13 Scheme of high-*Z* species migration by a multistep process of erosion – prompt re-deposition – re-erosion cycle.

5.6.2.2 Erosion under high-power loads and off-normal events

Fast events, such as disruptions, vertical displacements, and high-energy edge-localized modes, may lead to enhanced erosion and damage of PFCs [28, 45]. It happens when the target temperature rises to a value over the melting point or the sublimation threshold. Graphite and CFCs do not melt, but at temperatures exceeding 2500°C, sublimation of surface atoms starts. The threshold temperature is specific for each grade of carbon materials. The rate of carbon may be increased by radiation-enhanced sublimation [28]. In case of momentary and highly localized power deposition, exceeding 2.8–3 GW m^{-2}, brittle destruction may take place, involving ejection of debris (0.1–2 mm) from the subjected target. It has been carefully monitored under laboratory conditions using a high-power electron beam for irradiation [46, 47]. In tokamaks, the process has not been detected, though it cannot be excluded, but it was observed in a reversed field pinch where plasma edge control is difficult and power loads in disruptions are localized to small areas. Ejection of carbon debris and its penetration into plasma has been documented by a combination of spectroscopy, cameras, magnetic measurements, and microscopy [47, 48].

Metals under high power loads melt. The difference in the melting points (T_m) of Be and W is over 2000°C: T_m(Be) = 1287°C and T_m(W) = 3422°C. This also explains the selection of tungsten as the divertor material, though the tungsten could melt under reactor conditions, as has been shown in dedicated experiments with test limiters in TEXTOR [49] (see Section 5.7.1) and deliberately misaligned lamella of the bulk tungsten divertor in JET-ILW [50]. Surface melting of beryllium limiters has been observed in JET-ILW [51], and an example of a molten edge of the upper dump plate is shown in Fig. 5.5b. Under a high magnetic field, the melt layer moves and has a tendency to splash in the form of small droplets [49, 52, 53].

It should also be stressed that already below the melting point, the bulk structure of metals is modified by recrystallization, leading to the change of thermomechanical properties, especially thermal conductivity. Recrystallization of W takes place at 1400°C–1500°C, while for Be, it occurs at around 900°C. Exact temperatures are dependent on the metal grade.

A specific example of melt erosion is arcing, which occurs when electrical discharge is sparked. This can be caused either by imperfection on the surface, that is, a protruding piece that plays the role of a precursor for a unipolar arc [28, 54], or if an arc is formed between two adjacent parts. Figure 5.13 shows damage to a gold-coated stainless-steel mirror from JET [55] as an example. Macroscopic damage to the coating is shown in Fig. 5.14a, while optical microscopy images in Fig. 5.14b–d document the structure of propagating arc tracks and molten areas.

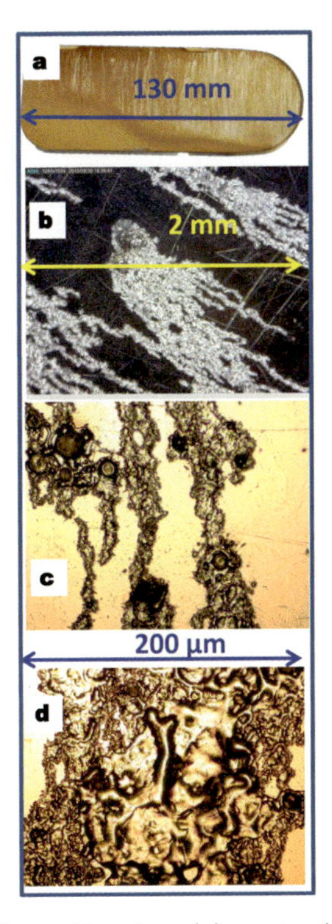

Figure 5.14 Damaged diagnostic mirror (a); arc tracks and melt zones (b–d).

5.6.2.3 Neutron-induced effects

Neutrons pass the armor and structural materials of the blanket, causing a volumetric change in material properties—in the composition by transmutation and in the structure by radiation damage to the lattice, such as by the formation of vacancies and interstitials. Bibliography on involved processes and their consequences is comprehensive; reviews can be found in Refs. [14, 56–59]. The direct impact on the modification of PFC surfaces is small, taking into account the energy of neutrons from reactions 5.1b and 5.2. However, changes induced in the bulk severely degrade the thermal conductivity [60] of PFCs, thus influencing the cooling efficiency, because the heat diffusion time from the surface to the cooling pipe is increased. This, in turn, would lead to the overheating of plasma-facing surfaces and then even to the damage of entire components.

5.7 Tools for Material Migration Studies

The list of diagnostic tools for plasma edge and wall studies is long. It stretches from optical spectroscopy and mass spectrometry to electrical instruments (Langmuir and retarding field analyzer [RFA]) and various erosion-deposition probes (EDPs). In principle, nearly every piece that can be retrieved from a tokamak or a stellarator after a short exposure (seconds) or a long-term campaign (many hours of plasma operation) may be a source of a certain type of valuable information, provided that the history of exposure is well known.

5.7.1 Erosion-Deposition Probes and Test Limiters

This section is focused on EDPs exposed to plasma and then analyzed mainly ex situ, while in situ techniques based on spectroscopy have been presented in Ref. [19]. Tools for erosion-deposition studies belong to the following basic categories: passive and active probes and marker and/or instrumented tiles. Over the years, many categories of EDPs have been used in fusion devices, including TEXTOR [61] and JET [62].

A category of passive probes incorporates various deposition monitors: collector probes for time-resolved and time-integrated measurements of particle fluxes, cavity-type traps for neutrals, and dust collectors and other types of surfaces installed in different places in the machine. The last category comprises the so-called wall inserts, louver clips, and also diagnostic mirrors; their specific application, construction, and location in JET can be found in Ref. [62]. All of them are placed in the SOL, that is, outside the LCFS. Only fast reciprocating probes are designed to go inside the plasma for a very limited period of time, the total cycle lasting 100 ms or less. They are for electrical measurements, but surface analyses of the probe heads deliver information about impurity fluxes. Three such heads used at JET before year 2009 are shown in Fig. 5.15a–c: an retarding field analyser (RFA), a turbulent transport multipin probe, and a collector combined with Langmuir, respectively. Another example of an active erosion-deposition probe is represented by a quartz microbalance (QMB) device equipped with a shutter, which is open only during selected discharges [63].

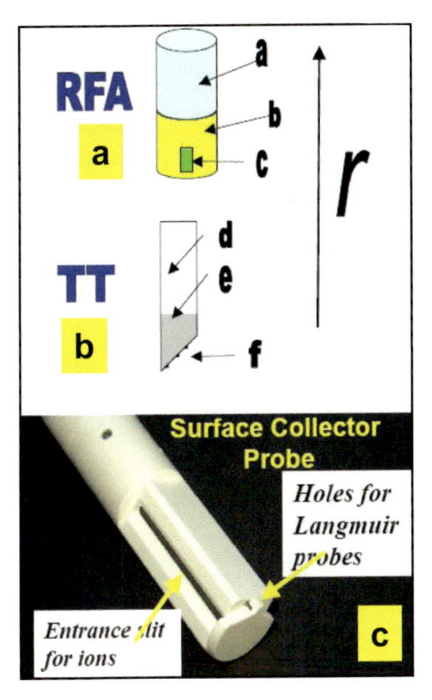

Figure 5.15 JET-C. Probe heads used both for electrical measurements and as collector probes: turbulent transport (a); retarding field analyzer (b); Langmuir and collector combined (c).

A family of marker tiles comprises a big range of limiter or divertor plates/blocks coated with several layers of marker films. The change in the layer thickness informs about the net deposition or the net erosion. Instrumented tiles are even more complex. They are equipped with a gas inlet system for the localized and active modification of the plasma edge composition and the material surface. Active modification is also realized by laser pulse ablating species from the tested material surface. Laser-assisted techniques allow for in situ studies of surfaces, for example, the removal of deposits. Secondly, experiments performed under diagnosed (i.e., controlled as much as possible) conditions have given invaluable inputs for advanced computer modeling and thus allowed for deep insight into the erosion and deposition processes. As a result of that integrated experimental and modeling approach, the understanding and assessment of behavior of various materials and their impact on the tokamak operation have been improved. It must be stressed that they were well diagnosed during exposure to the plasma. A spectroscopy arrangement for the observation and diagnosis of test limiters in TEXTOR is shown in Fig. 5.16.

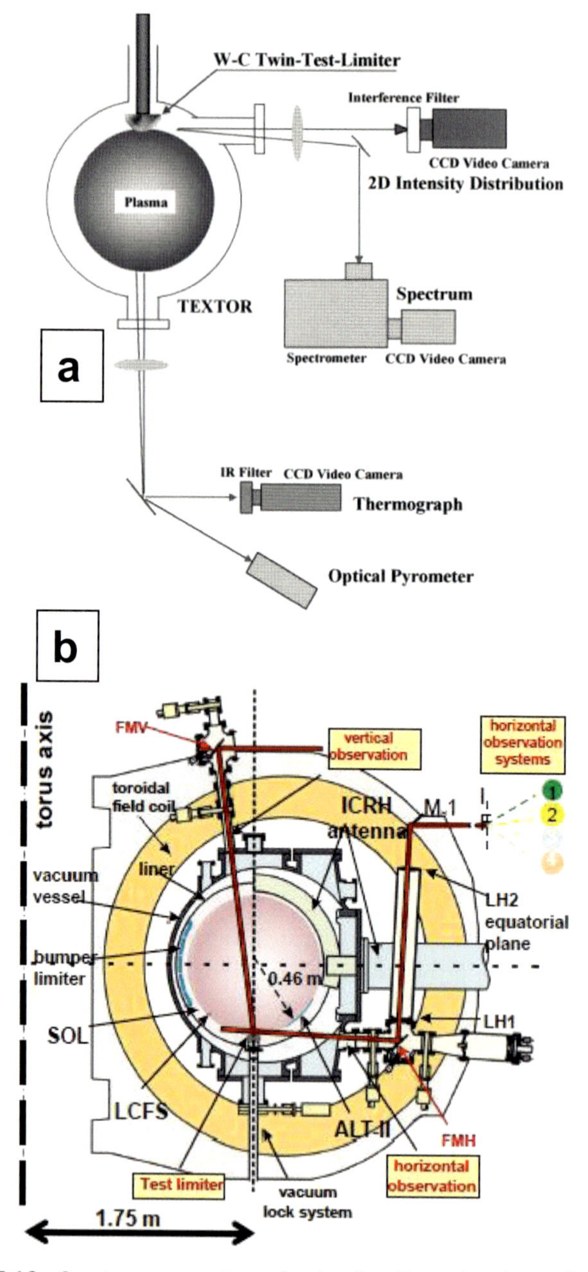

Figure 5.16 Spectroscopy systems for in situ diagnosis of test limiters in TEXTOR: upper limiter lock (a) and the bottom one (b).

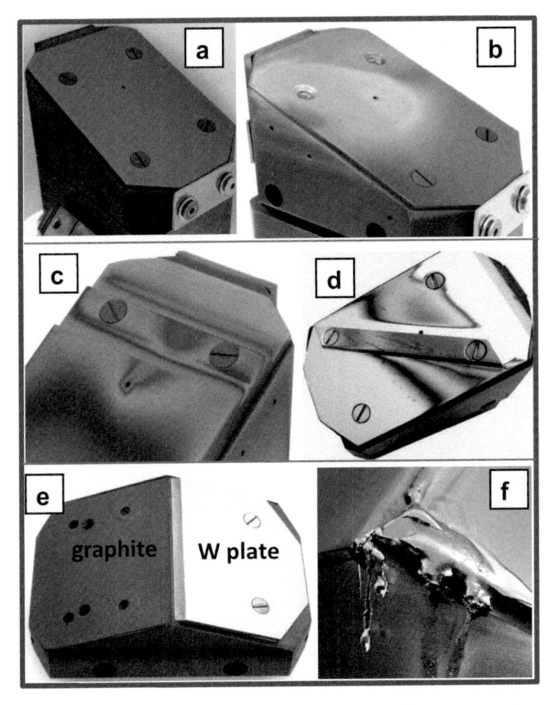

Figure 5.17 TEXTOR. Roof-shaped test limiters acting as erosion-deposition probes in studies of local deposition in high-*Z* gas (WF_6) injection experiments: a block before exposure (a) and after exposure, with deposition pattern (b); systems used for studies of deposition in shadowed areas (c, d); a setup for tungsten melting studies: as-installed W plate (e) and with a molten edge after exposure (f).

A comprehensive overview of probes used at TEXTOR has been presented in Ref. [61]. There have been more than a hundred different probes and test limiters exposed. Such broad research capabilities were offered by three transfer systems for probes. This included two so-called limiter locks, where heating, cooling, and local diagnosis of tested limiter was possible [19]. Over the years the transfer systems have been used to carry out (i) the performance test of high-*Z* metals and carbon-based composites considered as possible candidates for PFCs in next-step devices, (ii) material transport studies, and (iii) laser-assisted and gas puff–assisted modification of surfaces and the edge plasma. Images in Figs. 5.17 and 5.18 show a variety of test limiters: roof and mushroom shaped. Structures in Fig. 5.17a–d were mainly designed for local deposition studies in gas injection experiments and in the assessment of enhanced deposition in the so-called shadowed areas, that is, places without a direct plasma line of sight. The appearance of a limiter before the injection and after the experiment is shown in Figs. 5.17a and 5.17b, respectively, while the deposition in the shadowed region is documented in Figs. 5.17c and 5.17d. Controlled tungsten melting and melt layer motion was the aim of the experiment illustrated in Figs. 5.17e and 5.17f [49].

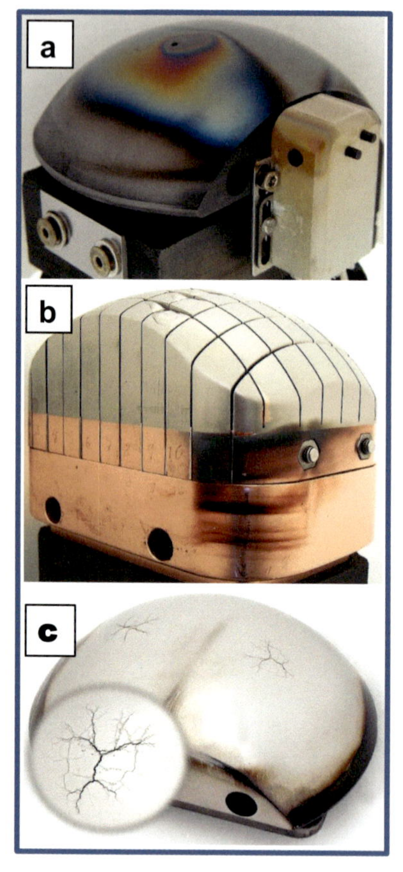

Figure 5.18 TEXTOR. Mushroom-type test limiters: the local deposition pattern after tracer experiments with $^{13}CH_4$ (a); the castellated tungsten test limiter (b) and the tungsten block after a sacrificial experiment with high power loading: the cracked metal is visible (c).

In the upper frame of Fig. 5.18 there is a graphite limiter with a hole for injecting gases during a discharge. This was used for local transport studies of injected tracers, mainly $^{13}CH_4$; tracers are described in the next section. One perceives a deposition pattern around the hole. Data from spectroscopy during the discharge and ex situ surface analyses were crucial for benchmarking transport codes. In Figs. 5.18b and 5.18c, there are tungsten limiters: a castellated structure and a solid one. The latter was damaged as a result of a high thermal shock. A network of macrocracks is clearly visible after the sacrificial type of experiment aiming at the assessment of limits in power handling by a component.

5.7.2 Tracer Techniques

The term "tracer" denotes an alien agent introduced deliberately into a studied system in minute quantities. The introduced agents must be reasonably harmless to the system under examination. Tracers are applied in many fields of science and technology in order to conclusively determine a reaction mechanism, distribution of components, flow direction, and velocity, that is, to reveal decisive steps in studied processes. An important point in using tracers in CFDs is the determination of long-range (i.e., meters) and short-range (local, i.e., in centimeters) material transport in the plasma. Such techniques have been used to study material migration, which is decisive for the lifetime of PFCs and fuel inventory. A tracer is introduced on purpose into the plasma edge, by puffing exotic (e.g., WF_6 and MoF_6) or isotope-labeled (e.g., $^{13}CH_4$ and $^{15}N_2$, $^{18}O_2$) gases, by ablating tracer material using lasers, or by exposing marker tiles coated with sandwich-type layers of heavy and light elements. An experimental approach to the determination of erosion phenomena involves the combined application of spectroscopy, mass spectrometry, surface probes for ex situ studies, and tracer materials. The selection and application of tracers requires the availability of a tracer and its affordability, proper surface probes placed in several locations, and relevant analytical methods in order to obtain a deposition pattern.

For many years a program dedicated to testing PFCs and the study of erosion processes by tracer techniques has been carried out at the TEXTOR tokamak, especially using methane labeled with carbon-13 [64]. This was followed by ^{13}C migration experiments in JET [23] and also in other machines [65]. The aim of the study was to determine carbon transport to the shadowed areas, where the greatest co-deposition of fuel was measured. In the case of JET, these were shadowed areas in the inner divertor.

Another example is related to the determination of nitrogen retention in PFCs. Nitrogen is injected for plasma edge cooling by radiation. The use of injected radiators is considered for ITER to cool the plasma edge and, by this, to avoid excessive heat loads on the divertor plates [66, 67]. Nitrogen, together with noble gases (Ne, Ar, Kr), is a candidate for edge radiators. One concern associated with the use of nitrogen is its in-vessel retention by implantation, co-deposition, or compound formation with wall materials [68]. The release of large amounts of accumulated nitrogen negatively affects the subsequent plasma operation. The assessment of nitrogen retention by ex situ analysis is difficult because one cannot fully discriminate between species retained in the exposure to plasma and those adsorbed on the surface on contact with air during the probe transfer to a surface analysis station. For nitrogen, the rare isotope ^{15}N (natural isotopic abundance 0.37%) is used. Enriched gas (>99% $^{15}N_2$) was used instead of N-14 in dedicated experiments performed in the TEXTOR and AUG tokamaks [69]. After this proof of principle, a series of experiments with the N-15 tracer have then been carried out in JET [70]. These studies have been facilitated by the development of analytical tools for surface analysis of N-15, as described in Section 5.8.1. A similar situation occurs when oxygen impurities are studied. Experiments with high-Z isotopic tracers have also been done in DIII-D [71].

5.8 Analysis of Wall Materials

A comprehensive approach to PWI studies comprises three fundamental elements:

- Experiments in tokamaks and relevant PWI simulators and material test facilities
- Ex situ and in situ analysis of wall components and probes
- Modeling

Therefore, analysis is not an isolated activity but an integral part of the entire research program. Its main role is to help understand processes that modify materials and lead to the degradation of their properties and to the contamination of fusion plasmas by species eroded from the wall. The analysis must provide data for the assessment of erosion-deposition pattern in the entire vessel and, by this, for modeling of material transport.

One needs answers to several fundamental questions: *what* has happened and *why* in order to plan *how* to deal with a given problem. The list of specific questions is broad as one deals with local and global aspects of material modification.

- Where are the erosion zones, and how are the materials modified by erosion and re-deposition?
- Where are the eroded species re-deposited, and how much fuel and which form is retained in the wall?
- What is the impact of neutrons on material properties and performance?
- How are the diagnostic components affected by PWI processes?
- What is the impact of wall conditioning and plasma edge cooling on PFC morphology?

5.8.1 Analyzed Species

The focus of analyses is on the retention of fuel species (D and T) and on properties of the first wall materials, because the aim is to obtain a comprehensive overview of the migration pattern. This requires a properly selected set of wall components (e.g., limiter and divertor tiles), a set of tools such as erosion-deposition diagnostics (various wall probes, instrumented tiles), material transport tracers, and a number of specialized laboratories with advanced apparatus and the capability to handle reactor materials contaminated, for instance, by Be and T. The analyses are carried out for all types of species (materials, elements and their isotopes) deliberately or accidently installed in or inserted, introduced, or injected into the torus.

- Hydrogen isotopes: H, D, and T
- Helium isotopes: ^{4}He being a fusion product or used for wall conditioning and ^{3}He used for heating with radio frequency (RF) waves

- Major PFC materials: beryllium, tungsten, and carbon
- Constituents of the vacuum vessel wall: Fe, Ni, Cr, Mo, Nb, Hf, W from stainless steel, and Inconel® alloy
- Components of materials tested as candidates for PFC: C-B, C-Si, and C-Ti composites
- Elements used for wall conditioning by evaporation or plasma-assisted film deposition: Li, Be, B, C, and Si
- Gases seeded for plasma edge cooling: N, Ne, Ar, Xe
- Species used in diagnostic beams: ^{6}Li and ^{7}Li
- Various markers used for material transport studies: For example, ^{10}Be, ^{10}B, ^{11}B, ^{13}C, ^{15}N, ^{18}O, ^{21}Ne, ^{22}Ne
- Fluorine from gaseous carriers of high-*Z* metals, WF_6 or MoF_6, used as markers in transport studies or tested for in situ repair of metal coatings
- Laser-ablated markers (e.g., Hf) and components of sandwich-type marker tiles: B, Re, Ta

Rich photographic documentation of PFCs in fusion devices included in this work allows for better understanding of analytical challenges. The shapes, sizes, and weights of components and plasma-induced surface features are to be taken into account because they have a serious impact on laboratory procedures. Colorful patterns prove a nonuniform composition attributed to erosion-deposition processes. As a consequence, there is a need for mapping the distribution of various species over large surfaces. This, in turn, calls for surface analysis stations with large chambers, robust sample holders, and precise manipulators of long travel [72, 73] to avoid sectioning (breaking or cutting) of tiles/specimens/ probes if possible. The decision to section or not is related to a range of practical aspects: (i) the cost and unique character of components, which—in many cases—may or must be reinstalled in fusion devices after ex situ sampling and/or analysis, (ii) research needs, which include certain types of necessary studies; for example, metallography requires cutting, polishing, and etching, while for thermal desorption, small samples are needed in most cases, and (iii) reduction of the contamination level to be handled in a laboratory [74, 75].

5.8.2 Analysis Methods

Nearly 50 different techniques have been used to respond to such comprehensive research needs. This has provided both fundamental and specific information on the morphology of targets exposed to a plasma. It should be stressed that the development of analytical tools takes place both at academic institutions [76] and at specialized industrial companies.

All methods are based on the same principle: sending a probing signal (signal in) and receiving a response from the studied matter (signal out). Various probing and outgoing generated signals are particles (ions, electrons, neutrons, neutral atoms, or molecules), photons of a very broad energy spectrum (γ, X-ray, to RF), sound, and magnetic or electric field. A single probing medium/beam can generate a wide range of signals leaving the matter under examination. For instance, irradiation with ions may lead to the emission of ions and neutrals (monoatomic or molecular), scattering of primary ions, generation of recoiled particles, photons originating from electronic and nuclear excitations, and a variety of nuclear reaction products, including neutrons. Taking into account a broad energy range (a few electron volts to millions of electron volts) and various types of primary beams (e.g., H^+, $^{3}He^+$, $^{12}C^{3+}$, $^{14}Si^{4+}$, $^{127}I^{9+}$), one gets a huge number of combinations, creating a palette of research possibilities.

Particular aims of the analyses are to determine material morphology and its changes under the plasma impact: structure (surface and bulk) and composition (elemental, isotopic, and chemical). There is no single method capable of addressing all these points. Each method has advantages but also serious limitations or even certain types of drawbacks, that is, destructive character. One selects a proper set of tools to solve a given problem, in other words, a set of most efficient techniques. A review of techniques has already been given in earlier articles [72, 73, 77]. The term "most efficient" denotes methods for the sensitive and selective determination of the content and distribution (lateral and in-depth) of hydrogen isotopes and several low-, medium-, and high-Z elements listed above. High-speed analysis and capabilities of probing hundreds of points over large areas of PFCs is another desired feature. Such criteria are met by ion beam analysis (IBA) methods, especially accelerator-based techniques [78, 79]. The principle of IBA is the irradiation of a solid with a monochromatic collimated ion beam followed by energy and/or mass analysis of species leaving the target. The frequently used methods are Rutherford backscattering spectroscopy (RBS), a large family of options offered by NRA, and particle-induced X-ray emission (PIXE). Measurements are carried out with a standard beam (diameter 0.8 mm) or a microbeam (5–8 μm size: μ-RBS, μ-NRA, μ-PIXE), depending on the available equipment and research needs. For special tasks one uses accelerator mass spectrometry and time-of-flight high-energy elastic recoil detection. Research capabilities of various accelerator laboratories are in Ref. [73], while conditions for the practical accelerator-based IBA of reactor materials have been compiled in Table 1 of Ref. [72]. That table, however, should be treated only as an indication, not as a carrier of ultimate recommendation. There is also continuous progress in the IBA of reactor materials, for instance, fuel retention analysis with high-energy ^{3}He [80] and development of detectors [76].

A second set of indispensable methods is connected with microscopy. Detailed surface features of PFCs (erosion and deposition areas with rough or flaking co-deposits) and dust particles have been presented on many occasions and also in this work (e.g., Figs. 5.8 and 5.10). Progress in the examination of near-surface structures of PFC surfaces and dust particles is connected with the application of the focused ion beam technique combined with ultra-high-resolution microscopy, especially scanning transmission electron microscopy [81, 82].

A crucial point to be taken into account in studies of materials retrieved from tokamaks is contamination by radioactive substances (tritium or activation/transmutation products) and—in the case JET—additionally by beryllium. These issues have been reviewed in Ref. [83], where work procedures in the Beryllium Handling Facility at JET are also explained.

In summary, *conditio sine qua non* for conclusive studies is in the selection of a proper set of techniques for solving a given problem. The next step is to have the chosen methods "at reach." There is no single laboratory or even a research center having all of them. The answer is simple: clear definition of a task and broad cooperation. It is also stressed that analysis must meet contemporary research requirements. In turn, advances in analysis widen experimental capabilities and plans.

5.9 Concluding Remarks

A brief overview provided in the chapter can serve only as an introduction to the vast field of plasma-induced modification of materials surrounding such an aggressive environment confined to a small volume of the vacuum vessel. Temperature gradients between the hot plasma and the cold wall belong to the greatest in the universe known to us. One talks about challenges but also about solutions in plasma edge engineering for a better understanding and control of interactions. For those interested in further studies on the topic, one can recommend overviews placed in the list of references, including also a coherent collection of tutorials from topical summer schools, for example, Ref. [84].

Acknowledgments

This work has been carried out within the framework of the EUROfusion Consortium and has received funding from the Euratom Research and Training Programme 2014–2018 and 2019–2020 under Grant Agreement No. 633053. The views and opinions expressed herein do not necessarily reflect those of the European Commission.

References

1. H.-S. Bosch, G. M. Halle, Improved formulas for fusion cross-sections and thermal reactivities, *Nucl. Fusion*, **32** (1992) 611.
2. J. Wesson, *Tokamaks*, 3rd ed., Oxford Science Publications, Clarendon Press, Oxford (2004).
3. E. Moses, Advances in inertial confinement fusion at the National Ignition Facility (NIF), *Fusion Eng. Des.*, **85** (2010) 983.
4. I. Hoffman, Review of accelerator driven heavy ion nuclear fusion, *Matter Radiat. Extremes*, **3** (2018) 1.

5. L. A. Artsimovitsch, Tokamak devises, *Nucl. Fusion*, **12** (1972) 215.
6. L. Spitzer, The stellarator concept, *Phys. Fluids*, **1** (1958) 53.
7. P.-H. Rebut, JET preliminary tritium experiment, *Plasma Phys. Control. Fusion*, **34** (1992) 1749.
8. D. Stork et al., Systems for the safe operation of the JET tokamak with tritium, *Fusion Eng. Des.*, **47** (1999) 131.
9. T. T. C. Jones et al., Technical and scientific aspects of the JET trace-tritium experimental campaign, *Fusion Sci. Technol.*, **48** (2005) 250.
10. C. H. Skinner, C. A. Gentile, J. C. Hosea, D. Mueller, J. P. Coad, G. Federici, R. Haange, Tritium experience in large tokamaks: application to ITER, *Nucl. Fusion*, **39** (1999).
11. D. Mueller et al., Tritium removal from TFTR, *J. Nucl. Mater.*, **241–243** (1997) 897.
12. T. Tanabe (ed.), *Tritium: Fuel of Fusion Reactors*, Springer Verlag (2016).
13. M. Kovari, M. Coleman, I. Cristescu, R. Smith, Tritium resources available for fusion reactors, *Nucl. Fusion*, **58** (2018) 026010.
14. M. Rubel, Fusion neutrons: tritium breeding and Impact on Wall Materials and components of diagnostic systems, *J. Fusion Energy*, **38** (2019) 315–329.
15. M. Enoeda et al., R&D status of water cooled ceramic breeder blanket technology, *Fusion Eng. Des.*, **89** (2014) 1131.
16. G. Aiello et al., Development of the helium cooled lithium lead blanket for DEMO, *Fusion Eng. Des.*, **89** (2014) 1444.
17. K. Ikeda, Progress in the ITER physics basis, *Nucl. Fusion*, **47** (2007).
18. N. Holtkamp, The status of the ITER design, *Fusion Eng. Des.*, **84** (2009) 98.
19. U. Samm, TEXTOR: a pioneering device for new concepts in plasma-wall interaction, exhaust, and confinement, *Fusion Sci. Technol.*, **47** (2005) 73; TEXTOR special issue.
20. J. Winter, Carbonisation in tokamaks, *J. Nucl. Mater.*, **145–147** (1987) 131.
21. J. Winter et al., Boronisation in TEXTOR, *J. Nucl. Mater.*, **162–163** (1989) 713.
22. J. Winter et al., Improved plasma performance in TEXTOR with silicon coated surfaces, *Phys. Rev. Lett.*, **71** (1993) 149.
23. J. P. Coad et al., Material migration and fuel retention studies during the JET carbon divertor campaigns, *Fusion Eng. Des.*, **138** (2019) 78.

24. G. F. Matthews et al., JET ITER-like wall: overview and experimental programme, *Phys. Scr.*, **T145** 014001 (2012).
25. Bosch et al., Technical challenges in the construction of the steady-state stellarator Wendelstein 7-X, *Nucl. Fusion*, **53** (2013) 126001.
26. T. Klinger et al., Performance and properties of the first plasmas of Wendelstein 7-X, *Plasma Phys. Control. Fusion*, **59** (2017) 014018.
27. W. O. Hofer, J. Roth (eds.), *Physical Processes of the Interaction of Fusion Plasmas with Solids*, Academic Press, New York (1996).
28. G. Federici et al., Plasma-material interactions in current tokamaks and their implications for next step fusion reactors, *Nucl. Fusion*, **41** (2001) 1967.
29. V. Philipps, P. Wienhold, A. Kirschner, M. Rubel, Erosion and redeposition of wall material in controlled fusion devices, *Vacuum*, **67** (2002) 399.
30. G. F. Matthews, Plasma operation with an all metal first-wall: comparison of an ITER-Like Wall with a carbon wall in JET, *J. Nucl. Mater.*, **438** (2013) S2.
31. T. Loarer et al., Comparison of long term fuel retention in JET between carbon and the ITER-Like Wall, *J. Nucl. Mater.*, **438** (2013) S108.
32. S. Brezinsek et al., Fuel retention studies with the ITER-Like Wall in JET, *Nucl. Fusion*, **53** (2013) 083023.
33. M. Merola et al., Overview and status of ITER internal components, *Fusion Eng. Des.*, **89** (2014) 890.
34. P. Sigmund, Theory of sputtering. I. Sputtering yield of amorphous and polycrystalline targets, *Phys. Rev.*, **184** (1969) 383.
35. R. Behrisch, W. Eckstein (eds.), *Sputtering by Particle Bombardment: Experiments and Computer Calculations from Thresholds to MeV Energies*, Topics in Applied Physics Vol. 110, Springer Verlag (2007).
36. J. Roth et al., Recent analysis of key plasma-wall interaction parameters for ITER, *J. Nucl. Mater.*, **390–391** (2009) 1.
37. J. F. Ziegler, J. P. Biersack, M. D. Ziegler, *SRIM - The Stopping and Range of Ions in Matter*, SRIM Co., Chester (2008).
38. M. Rubel, P. Wienhold, D. Hildebrandt, Fuel accumulation in co-deposited layers on plasma-facing components, *J. Nucl. Mater.*, **290–293** (2001) 473–477.
39. M. Rubel, G. Sergienko, A. Kreter, A. Pospieszczyk, M. Psoda, E. Wessel, An overview of fuel retention and morphology in a castellated tungsten limiter, *Fusion Eng. Des.*, **83** (2008) 1049–1053.

40. W. Wolfram, *Gmelins Handbuch der Anorganischen Chemie*, System-Nummer 54, Verlag Chemie, Berlin (1933), p. 110.
41. M. Psoda, M. J. Rubel, G. Sergienko, P. Sundelin, A. Pospieszczyk, Material mixing on plasma –facing components: compound formation, *J. Nucl. Mater.*, **386–388** (2009) 740–743.
42. F. Koch, S. Köppl, H. Bolt, Self passivating W-based alloys as plasma-facing material, *J. Nucl. Mater.*, **386–388** (2009) 572.
43. D. Naujoks, R. Behrisch, Erosion and re-deposition at the vessel walls in fusion devices, *J. Nucl. Mater.*, **220–222** (1995) 227.
44. M. Rubel et al., Tungsten migration studies by controlled injection of volatile compounds, *J. Nucl. Mater.*, **438** (2013) S170
45. A. Loarte et al., Chapter 4: Power and particle control, *Nucl. Fusion*, **47** (2007) S203.
46. H. Bolt, J. Linke, H. J. Penkalla, E. Tarret, Emission of solid particles from carbon materials under pulsed surface heat loads, *Phys. Scr.*, **T81** (1999) 94.
47. J. Linke et al., Emission of carbon particles, brittle destruction and co-deposit formation: Experience from electron beam experiments and controlled fusion devices, *Phys. Scr.*, **T91** (2001) 36.
48. M. Rubel et al., Dust particles: Morphology, observation in the plasma and the influence on the plasma operation, *Nucl. Fusion*, **41** (2001) 1087.
49. G. Sergienko et al., Tungsten behaviour under power loads of fusion plasmas, *Phys. Scr.*, **T128** (2007) 81–86.
50. J. W. Coenen et al., Transient induced tungsten melting at the Joint European Torus (JET), *Phys. Scr.*, **T170** (2017) 014013.
51. I. Jepu et al., Beryllium melting and erosion on the upper dump plates in JET during three ILW campaigns, *Nucl. Fusion*, **59** (2019) 086009.
52. D. Ivanova et al., Survey of dust formed in the TEXTOR tokamak: structure and fuel retention, *Phys. Scr.*, **T138** (2009) 014025.
53. M. Rubel et al., Dust generation in tokamaks: overview of beryllium and tungsten dust characterisation in JET with the ITER-Like Wall, *Fusion Eng. Des.*, **136** (2018) 579–586.
54. F. Schwirzke, Unipolar arc model, *J. Nucl. Mater.*, **128–129** (1984) 609.
55. A. Garcia-Carrasco et al., Plasma impact on diagnostic mirrors in JET, *Nucl. Mater. Energy*, **12** (2017) 506.
56. M. R. Gilbert, J.-Ch. Sublet, Neutron-induced transmutation effects in W and W-alloys in a fusion environment, *Nucl. Fusion*, **51** (2011) 043005.

57. M. R. Gilbert et al., An integrated model for materials in a fusion power plant: transmutation, gas production, and helium embrittlement under neutron irradiation, *Nucl. Fusion*, **52** (2012) 083019.

58. N. Baluc, Materials for fusion power reactors, *Plasma Phys. Control. Fusion*, **48** (2006) B165.

59. Y. Wu, *Fusion Neutronics*, Springer Verlag (2017).

60. J. Linke et al., Emission of carbon particles, brittle destruction and co-deposit formation: Experience from electron beam experiments and controlled fusion devices, *Phys. Scr.*, **T91** (2001) 36.

61. M. Rubel et al., Overview of wall probes for erosion and deposition studies in the TEXTOR tokamak, *Matter Radiat. Extremes*, **2** (2017) 87–104.

62. M. Rubel et al., Overview of erosion–deposition diagnostic tools for the ITER-Like Wall in the JET tokamak, *J. Nucl. Mater.*, **438** (2013) S1204.

63. A. Kreter et al., Nonlinear impact of edge localized modes on carbon erosion in the divertor of the JET tokamak, *Phys. Rev. Lett.*, **102** (2008) 045007.

64. P. Wienhold et al., Investigation of carbon transport in the scrape-off layer of TEXTOR-94, *J. Nucl. Mater.*, **290–293** (2001) 362–366.

65. S. L. Allen et al., C-13 transport studies in L-mode divertor plasmas on DIII-D, *J. Nucl. Mater.*, **337–339** (2005) 30.

66. S. Brezinsek et al., Beryllium migration in JET ITER-Like Wall plasmas, *Nucl. Fusion*, **55** (2015) 063021.

67. A. Kallenbach et al., Plasma surface interactions in impurity seeded plasmas, *J. Nucl. Mater.*, **415** (2011) S19.

68. M. Rubel et al., Nitrogen and neon retention in plasma-facing materials, *J. Nucl. Mater.*, **415** (2011) S223.

69. P. Petersson et al., Development and results of global tracer-injection technique with ^{15}N in ASDEX upgrade, *J. Nucl. Mater.*, **438** (2013) S616.

70. P. Petersson et al., Co-deposited layers in the divertor region of JET-ILW, *J. Nucl Mater.*, **463** (2015) 814.

71. E. A. Unterberg et al., Use of isotopic tungsten tracers and a stable isotope-mixing model to characterize divertor source location in the DIII-D metal rings campaign, *Nucl. Mater. Energy*, **19** (2019) 358.

72. M. Rubel, Analysis of plasma facing materials: material migration and fuel retention, *Phys. Scr.*, **T123** (2006) 54.

73. M. Mayer et al., Ion beam analysis of fusion plasma-facing components and materials: facilities and research challenges, *Nucl. Fusion*, **60** (2020) 025001.

74. R. D. Penzhorn et al., Tritium depth profiles in graphite and carbon fibre composite material exposed to tokamak plasmas, *J. Nucl. Mater.*, **288** (2001) 170.

75. M. Rubel et al., Beryllium and carbon films in JET following D-T operation, *J. Nucl. Mater.*, **313–316** (2003) 321–326.

76. P. Ström, P. Petersson, M. Rubel, G. Possnert, A combined segmented anode gas ionization chamber and time-of-flight detector for heavy ion elastic recoil detection analysis, *Rev. Sci. Instrum.*, **87** (2016) 103303.

77. M. Rubel et al., The role and application of ion beam analysis for studies of plasma-facing components in controlled fusion devices, *Nucl. Instrum. Methods*, **B371** (2016) 4.

78. W. K. Chu, M. Mayer, M. A. Nicolet, *Backscattering Spectroscopy*, Academic Press, New York (1978).

79. J. R. Tesmer, M. Nastasi (eds.), *Handbook of Modern Ion Beam Analysis*, Material Research Society, Pittsburg, PA (1995).

80. M. Mayer et al., Quantitative depth profiling of deuterium up to very large depths, *Nucl. Instrum. Methods B*, **267** (2009) 506.

81. A. Baron-Wiechec et al., First dust study in JET with the ITER-Like Wall: sampling, analysis and classification, *Nucl. Fusion*, **55** (2015) 113033.

82. E. Fortuna-Zalesna et al., Studies of dust from JET with the ITER-Like Wall: composition and internal structure, *Nucl. Mater. Energy*, **12** (2017) 582.

83. A. Widdowson et al., Experience in handling beryllium, tritium and activated components from JET ITER-Like Wall, *Phys. Scr.*, **T167** (2016) 014057.

84. R. Koch, Ninth Carolus Magnus Summer School on plasma and fusion energy physics, *Fusion Sci. Technol.*, **57** T2 (2010) 185.

Chapter 6

Plasma-Assisted Wall Conditioning of Fusion Devices: A Review

F. L. Tabarés,[a] D. Tafalla,[a] and T. Wauters[b]
[a]*Laboratorio Nacional de Fusion, CIEMAT, Av Complutense 40, 28040 Madrid, Spain*
[b]*Laboratory for Plasma Physics, LPP-ERM/KMS, B-1000 Brussels, Belgium, Trilateral Euregio Cluster (TEC) Partner*
tabares@ciemat.es

6.1 Introduction

A comprehensive introduction to magnetically confined fusion has been given in the previous chapter so that the reader goes through the concepts in that chapter before starting the present one. In brief, the production of 1 GW of fusion power in the form of neutron outflow conveys the extraction of >200 MW in the form of He ash. This power has to be properly handled by the materials surrounding the plasma, and the resulting plasma-wall interaction processes represent one of the major challenges on the route to commercial fusion nowadays.

Plasma Applications for Material Modification: From Microelectronics to Biological Materials
Edited by Francisco L. Tabarés

ISBN 978-981-4877-35-0 (Hardcover), 978-1-003-11920-3 (eBook)
www.jennystanford.com

Furthermore, in magnetically confined fusion plasmas, a strong nonlinear connection exists between the processes occurring at the wall, the properties of the plasma edge, and the properties of the main plasma. The ion and electron temperature profiles, the power flow in the plasma edge and its radiated fraction (radiative edge), and the concentration and penetration depth of neutral hydrogen atoms and of plasma impurities into the scrape-off layer (SOL) are examples of such strongly interlinked quantities. By means of complex mechanisms, they modify energy and particle transport, plasma confinement, and the performance of the discharge. High performance implies maximizing the fusion triple product $n_i T_i \tau_E$ of deuteron and triton density n_i, ion temperature T_i, and energy confinement time τ_E in a thermal plasma. Hence it is important to achieve high densities, to avoid the cooling of the plasma by excessive radiation of power and the dilution of the triton and deuterons by impurities, and to promote high-energy confinement, consistent with the requirements of energy removal and helium pumping.

The plasma edge must have the appropriate architecture in order to obtain a thermonuclear fusion plasma with high performance. The edge plasma itself is strongly influenced by the plasma-wall interaction. The plasma-surface interaction processes have been thoroughly reviewed by several authors [1–3]. Several reviews of plasma-edge phenomena have also been published [4, 5], and previous reviews on first wall conditioning by cold plasma can be found in the literature [6–8].

Atoms and ions from the plasma have a strong interaction with matter. Their range is limited to typically several tens of nanometers from the surface even in low-Z target materials (Be or C) with small stopping powers. It is in this thin layer, the near-surface region, where plasma-surface interactions take place and where all the basic processes, like sputtering, chemical erosion, trapping, and reflection, occur. These processes may induce the release of wall material and thus introduce plasma impurities or have an effect on the hydrogenic particle balance (hydrogen recycling). It is obvious that the modification of this near-surface region, for example, by discharge cleaning or by depositing thin films of appropriate chemical composition and physical structure, will bring about major changes in the plasma-surface interaction processes and may thus allow active control of the phenomena critical to the plasma and its performance.

The role of the first wall in plasma-material interactions in a fusion plasma is in principle less relevant than that taking place in the materials directly exposed, such as limiters and divertor plates. However, the larger areas involved may sometimes compensate for that. A well-known example is the flux of high-energy neutral particles created by charge exchange processes between the energetic protons and the neutral atoms surrounding the plasma, at least for limiter configurations. Roughly half of the total impurity content of the plasma can be ascribed to the sputtering of the first wall by these energetic particles. Furthermore, thermal excursions of the first wall materials during long pulse operation generate the desorption of weakly bound molecular species, such as water and CO, which represent the major source of high recycling atomic oxygen in the plasma. The addition of a getter coating (B, Be, Ti, Li) on the inner side of the vessel walls has a dramatic effect on the impurity level of the hot plasma. In a similar fashion, maintaining the isotope purity of the plasma is challenging due to the charge exchange processes referred to above, and devoted plasma techniques are needed in this direction too.

Measurements in TEXTOR have shown that the high-energy fraction of the reflected neutral atoms is smaller for a limiter of low *Z* than of that of a high-*Z* material [9]. This can be understood from the dependence of the energy transfer on the mass ratio in collisions of projectile and substrate atoms. Thus, choosing low-*Z* wall elements appears to be favorable for minimizing the penetration of reflected neutrals. As suggested for the H-mode, but probably applying to other high-confinement regimes also, important ingredients in the recipes for achieving high confinement regimes appear to be low neutral hydrogen influxes, controlled impurity influxes (concerning both the species and the rates), and low-*Z* wall materials.

This review gives an overview of plasma-mediated wall conditioning techniques for the active control of two critical elements for high-performance operation: plasma impurities and hydrogen recycling. It must be pointed out here that other, simpler techniques are routinely used at earlier stages of the conditioning process. The most conspicuous one is baking the full-vacuum chamber, a rather tedious and limited process aimed at the removal of weakly bound molecular species, such as water, hydrocarbons, and carbon oxides. The presence of magnetic coils and delicate instrumentation, however, limits the upper temperature allowed in the thermal cycle and hence its efficiency. Thermal expansion of the vessel may become an important issue, especially when tolerances in some critical components are very low, as is the case for some stellarator configurations. The huge mass of the vacuum vessel calls for slow ramping of the temperature and long cooling times. Thus only in very exceptional conditions is this basic technique applied to fusion devices. The other well-known technique is film evaporation, titanium gettering being particularly spread out within the high-vacuum community. As low-Z elements are preferred as possible plasma contaminants, Be and Li substitute Ti as gettering film. Although they are not plasma-mediated conditioning techniques, the ionization of the evaporated atoms into a low-temperature plasma is sometimes promoted in order to enhance film homogeneity, a key issue in the achievement of long-lasting effects of the deposited films.

6.2 Fundamentals of Wall Conditioning by Plasmas

Regular machine operation of a fusion device is devoted to the generation of a fully ionized plasma, with electron temperatures on the order of hundreds to thousands of electron volts. Under such conditions, the atom concentration/ion concentration ratio is negligible. On the contrary, conditioning plasmas focus on chemically induced surface modifications and, therefore, high temperatures are generally avoided. One exception of this would be the use of hot, ohmic plasmas for the conditioning of the walls after a disruption, an event on which most of the energy content of the plasma is suddenly deposited on the surrounding material as a consequence of quickly evolving magnetohydrodynamic instabilities [10]. Such a scenario represents the most dangerous case for the machine integrity, and several techniques were developed to mitigate them as much as possible. Interestingly, regular wall conditioning is one of the most effective ways to prevent disruptions.

From a conceptual point of view, the underlying processes during plasma conditioning of the reactor walls are the same as those described in Chapter 5 in relation to the plasma-wall interaction phenomena, that is, physical and chemical sputtering, ion implantation, particle- and temperature-induced desorption, etc. While there is little room for control of these processes during hot plasma operation, as described in Chapter 5, they can be tailored in conditioning plasmas in order to optimize their efficiency. Let's take a look of some examples of that.

The selection of the gas on which the cold plasma is generated is of paramount relevance. Technical metals are known to have a score of light elements, like oxygen, carbon, and nitrogen, incorporated during their manufacturing. All these elements are chemically active against atomic hydrogen, leading to the formation of volatile molecules prone to removal by conventional vacuum pumping techniques. In the case of oxygen, one of the most conspicuous impurities, its removal through water formation implies wall temperatures high enough to prevent condensation, but ammonia- and methane-mediated release of N and C impurities, respectively, can be performed at or slightly above room temperature. The use of molecular oxygen or nitrogen as the plasma species is also a way to promote surface chemistry with similar effects, although ending the cleaning protocol with a hydrogen plasma is the only way to guarantee a contamination-free metallic surface suitable for hot plasma operation.

The molecular species created during the plasma-activated reactions will be partially decomposed by the conditioning plasma and, depending on the energy of the resulting fragments, reimplanted or trapped into the wall. Therefore, some tuning of the microscopic plasma parameters is deemed necessary.

A simple zero-dimensional particle balance of the processes described above reads as follows:

$$\frac{\mathrm{d}P}{\mathrm{d}t} = \gamma \cdot \Gamma_i \cdot [X] \cdot A - P \cdot k_\mathrm{d} \cdot n_\mathrm{e} - \frac{P}{\tau_\mathrm{R}}, \qquad (6.1)$$

where P stands for the molecular product of the reaction, of yield = γ, between the impinging species (H atoms and H_2^+ ions in a H_2 plasma) of flux = Γ_i (s^{-1}) and the surface contaminant, X, whose concentration in atoms/m^2 is $[X]$; A (m^2) stands for the plasma-wetted area; k_d ($m^3 \cdot s^{-1}$) represents the rate constant of decomposition by electron impact of P; n_e represents the electron density of the plasma (m^{-3}); and τ_R stands for the residence time of the molecule in the vacuum vessel of volume = V (m^3) and pumping speed = S (m^3/s); $\tau_R = V/S$. Under steady-state conditions, the concentration of the molecular product in the conditioning plasma is simply given by

$$[P] \cdot V = \frac{\gamma \cdot \Gamma_i \cdot [X] \cdot A}{k_d \cdot n_e + \frac{1}{\tau_R}}. \quad (6.2)$$

The microscopic parameters of the conditioning plasma, n_e and T_e, affect the removal rate of impurities through the vacuum system, $[P] \cdot S_{eff}$, in a combined way with the vacuum parameters (i.e., residence time). First, the dissociation/ionization rate constants of molecular species are very sensitive to the electron temperature in the range of a few electron volts (see Fig. 6.1) and the flux of the active species impinging on the walls is roughly proportional to its electron density. Since fusion devices are generally large vacuum vessels with restricted pumping access, rather long values of τ_R apply. Therefore, increasing Γ_i by increasing the plasma current (direct current [DC] glow discharge GD) or the injected power (microwave [MW] and radio frequency [RF]) conveys an increase in the plasma density as well and, according to Eq. 6.2, the rate of impurity removal becomes saturated. This problem is usually circumvented by using pulsed plasma sources and imposing plasma-off periods, that is, duty cycles, long enough to allow for the pumping down of the released molecular species (see Section 6.3).

So far, it has been assumed that the molecular species are directly released from the wall upon their formation by chemical reactions with the plasma, and first-order kinetics are implicit in Eq. 6.1. However, surface-specific phenomena may play a critical role in the conditioning process. The simplest of this phenomenon is adsorption, especially relevant for water-mediated removal of surface oxygen. This is only achieved as far as water desorbs from the surface once generated, calling for temperatures in the range of 100°C–150°C. When recombination of surface species is behind the cleaning mechanism, as N–H in the case of ammonia formation, a more complex dependence on plasma parameters than that implicit in Eq. 6.2 is observed. Furthermore, diffusion-limited kinetics are found in many instances, adding an extra temperature and time dependence to the release rate. Suitable models for impurity removal and particle recycling have been developed taking into account the surface-specific issues [11].

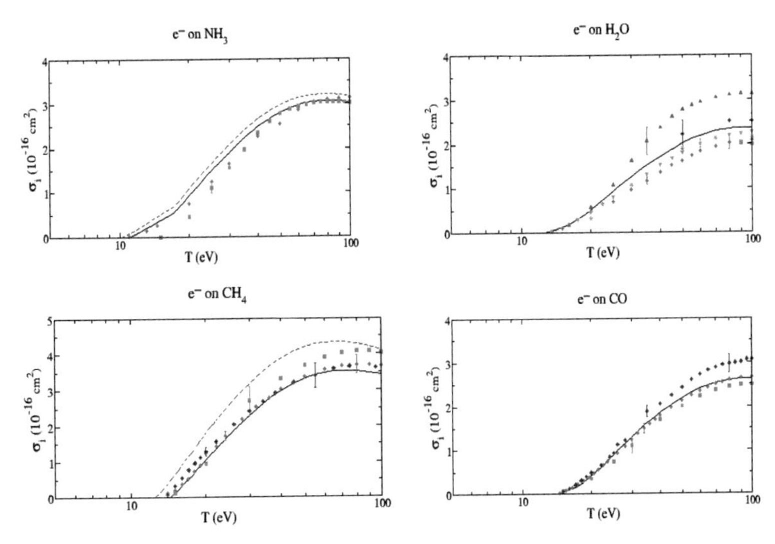

Figure 6.1 Ionization cross section of some molecular products of the plasma cleaning process for electron energies from the threshold to 100 eV (https://physics.nist.gov/PhysRefData/Ionization/molTable.html).

Aside from chemically based cleaning processes, noble gases are also commonly used in wall conditioning, especially He. Its use is particularly recommended for H inventory and isotope control, directly affecting recycling and plasma density control. Particle-induced recombination and desorption, leading to impurity release (e.g., CO formation), competes with physical sputtering, with a deleterious impact on windows and delicate internal instrumentation. It is sputtering avoidance that sets the optimal parameters for plasmas of this kind, such as the type of gas and the plasma potential. Another side effect reported on the use of He GD conditioning is plasma contamination by the implanted He, leading to density control issues. This is due to the high recycling of atomic He on metallic walls as well as its low solubility in the metal, eventually leading to microbubble formation. Tungsten walls are particularly prone to this effect. Pulsed instead of continuous He GD plasmas are applied in full W wall devices as Axially Symmetric Divertor Experiment (ASDEX) Upgrade [12].

As the inventory of impurities in the bulk of the material is very large, the effect of plasma cleaning of the topmost layer is only transient, and segregation from the bulk eventually pollutes the surface again after the conditioning period. The solution for a longer effect is plasma coating. Plasma-assisted deposition of low-*Z* coatings, such as C, B, or even Si, is a common practice in magnetic fusion research. Direct evaporation of the coating material is also used whenever possible. Such is the case of Be [13], Ti [14], and Li [15]. While evaporation under vacuum is restricted to line-of-sight deposition, leading to nonhomogeneous coatings, plasma-assisted techniques are in principle capable of producing evenly distributed films. However, as described in Chapter 5, erosion and redeposition phenomena during hot plasma operation define the final location of the deposited species. Real-time deposition by injection of the film precursor into the hot plasma is seen as an alternative to obtain steady coating conditions for long pulse fusion devices.

6.3 Kinds of Conditioning Plasmas

The kinds of plasmas used in fusion research for in situ wall conditioning are basically the same as those used in microelectronic manufacturing or in molecular dissociation (see Chapters 1 and 2). So, continuous (DC), RF, or MW cold plasmas have application in the field and can be used according to the needs and hardware availability. In the last two cases, some magnetic field is needed and resonant and nonresonant RF and MW waves can be launched through devoted antennas.

6.3.1 Direct Current Glow Discharge

This is by far the easiest and more used technique in fusion devices. Several anodes are introduced in the vacuum vessel, and the chamber is grounded in order to become the cathode of the discharge. Low pressures (10^{-3}–10^{-2} mbar) of the reactive gas are used in order to keep good vacuum pumping and to allow for the full development of the cathode fall so that energetic ions impinge on the walls. This requirement makes the ignition of the plasma rather challenging, and transient pressure rise or electron injection by a devoted gun is often used for a trouble-less breakdown. When available, MW- or RF-aided breakdown can be also used. Depending on the aim of the conditioning—chemically driven release of impurities or enhanced sputtering—the walls of the device are heated or kept at room temperature (see Section 6.2). In any case, mass spectrometry or monitor samples are used to follow the cleaning process in real time.

Figure 6.2 shows a typical image (top left) of DC He plasma applied in the TOMAS toroidal device at FZ-Jülich, Germany, using one calotte-shaped graphite anode located at the top of the vessel. The position and camera viewing angle are indicated in the figure on the right. GD plasma requires the absence of magnetic fields. The next images show the same discharge with the addition of a small toroidal magnetic field, increased in steps of about 1.2 mT. It is clearly seen that the glow current density becomes inhomogeneous due to the preferred transport of charged particles along magnetic field lines. Therefore, in superconducting devices like Wendelstein 7-X (W7-X), JT-60SA, and International Thermonuclear Experimental Reactor (ITER), where the toroidal magnetic field in the vessel is present for multiple hours or days continuously, the operation of glow discharge cleaning (GDC) is limited to machine shut-down periods.

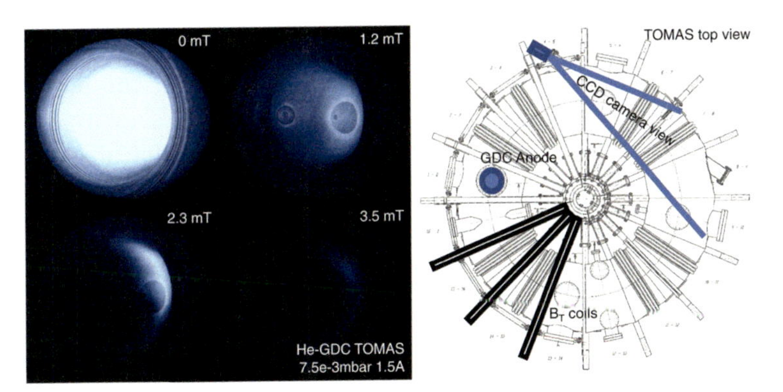

Figure 6.2 Tangential CCD images of a He-GDC plasma in TOMAS without toroidal magnetic field and with the application of small fields. The location of the anode with respect to the tangential camera view is indicated in the schematic top view of the device.

6.3.2 Conditioning Techniques in the Presence of Magnetic Fields

Chapter 5 introduced the two most promising magnetic confinement concepts: the tokamak and the stellarator. Plasma confinement in these devices relies on toroidally nested flux surfaces. The magnetic field topology follows from currents flowing inside and outside the plasma. A stellarator relies mostly on external currents that flow in complex shaped field coils. Hence, the nested flux surfaces are present also in the absence of plasma. A tokamak combines external planar coils with a toroidal current flowing inside the plasma to create the nested field line topology. Without this plasma current, only a toroidal field remains with greatly reduced plasma confinement.

Magnetized conditioning discharges in tokamaks or stellarators are, as their nonmagnetized counterpart, aimed at inducing a known and optimal flux of particles to the plasma-facing components (PFCs). The particle fluxes initiate the release of impurities and hydrogen isotopes or produce a thin surface coating. The preferred transport of charged particles along the magnetic field lines will, however, affect the intensity and properties of the conditioning flux for different wall components. This is illustrated in Fig. 6.3 for the typical surfaces in a fusion device. The figure includes a schematic representation of the dedicated components that are placed in the plasma vessel to receive and extract the heat load from the plasma. The interaction occurs at the inevitable intersection points of the outer magnetic field lines and the wall surface. At the divertor surfaces, the interaction is characterized by the shallow angle between the field lines and surface, spreading the ion flux along the field lines over a large surface area. At the limiter, on the other hand, the flux arrives mostly perpendicular to the surface. The main wall is typically a "shadowed" area, which receives a greatly reduced ion flux due to the strong plasma density decay behind the protection limiters. Neutrals, not affected by the magnetic field, can be the dominant flux component at such shadowed locations. This is also the case for the gaps between castellated structures where impurities are known to get codeposited with the plasma fuel. Neutrals can effectively condition such gaps.

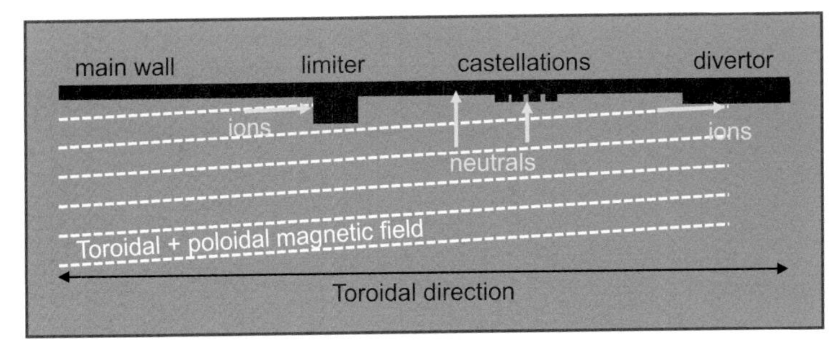

Figure 6.3 Illustration of the typical surfaces in a magnetized toroidal plasma device and their interaction with the plasma flux flowing along the magnetic field lines (white dashed lines).

Plasma surface interaction in a diverted plasma, for example, for conditioning purposes, is localized mostly at the divertor surfaces. The limiters in such a plasma are located in the far SOL with a much reduced plasma density. This is opposite to the RF and MW conditioning discharges in tokamaks (see Sections 6.3.3 and 6.3.4), where the limiters will receive the highest ion flux, while the flux at the divertor is being spread over its toroidal surface area.

In this overview of the different kinds of conditioning plasmas operated in the presence of a magnetic field, we categorize the techniques firstly by their magnetic field configuration. Secondly, because for a given field, the plasma can be produced by different methods, we separate (i) RF conditioning discharges produced by electromagnetic fields in the ion cyclotron range of frequencies (ICRF) on tokamaks and stellarators, (ii) MW conditioning on tokamaks by using electron cyclotron resonance (ECR) heating waves, and (iii) conditioning by high-temperature plasmas with nested flux surfaces, either EC plasma on a stellarator or ohmic discharges on a tokamak. An overview of these techniques is given in Table 6.1, together with the operation conditions and area of application. Finally, to illustrate that other magnetized discharge conditioning methods have been used in the past or will be discovered in the future, we introduce Taylor discharge cleaning, which played a role in the first superconducting fusion devices. More details on conditioning in superconducting devices can be found in Ref. [16].

Table 6.1 Overview of techniques for conditioning the plasma-facing components in three major superconducting fusion devices, tokamaks ITER (W/Be), JT-60SA (C), and stellarator W7-X (C), together with their typical pressure and electron density range

	B_T	Pressure (mbar)	Density (m^{-3})	SC device
DC GD conditioning	✓	10^{-3}–10^{-2}	$<10^{16}$	ITER JT-60SA W7-X
RF-ion cyclotron wall conditioning	✓	10^{-5}–10^{-4}	10^{16-18}	ITER W7-X
MW-electron cyclotron wall conditioning (tokamak)	✓	10^{-5}–10^{-4}	10^{18-19}	JT-60SA
HT-high-temperature-diverted plasmas (nested flux surfaces)	✓	$<10^{-5}$	10^{19-20}	ITER JT-60SA W7-X

6.3.3 RF Conditioning

Ion cyclotron wall conditioning (ICWC) was originally developed for stellarators and later successfully applied in tokamaks [16]. It uses the IC heating and current drive antenna systems (i) to initiate the discharge by the strong electric fields in the vicinity of the antenna and (ii) to maintain the steady homogeneous RF plasma by coupling ICRF power to both electrons and ions, mainly via nonresonant processes. The current-less discharge, when applied in a tokamak, is partially ionized with an electron density on the order of 1e17/m^3 and a temperature of about 5 eV. In stellarators, with better confinement, the discharge can be optimized for higher densities.

Figure 6.4a shows a tangential camera image of an ICWC discharge on tokamak JET at CCFE, UK, coupling 250 kW of IC power to the plasma at 25 MHz and a toroidal field of 3.3 T for a deuterium neutral pressure of 2e-5 mbar. The emitted plasma light in the visible range results mostly from collisional processes of electrons with background neutrals as well as with wall-released particles. The homogeneity of the plasma was achieved by applying a vertical magnetic field of 30 mT using poloidal field coils.

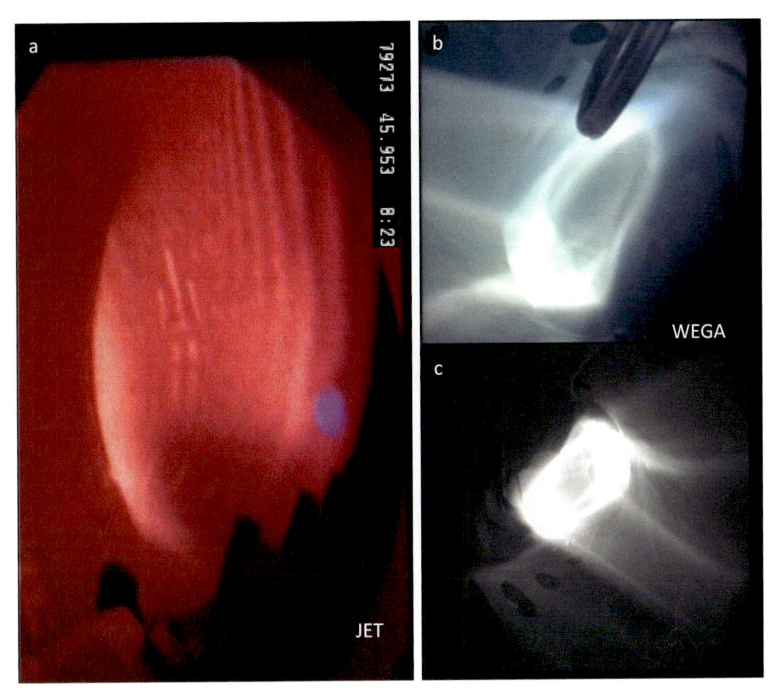

Figure 6.4 (a) CCD camera image of a hydrogen ICWC plasma in tokamak JET illustrating the homogeneous filling of the vacuum vessel by the RF plasma. Reprinted from Ref. [38], Copyright (2011), with permission from Elsevier. (b, c) CCD camera image of a helium ICWC plasma in stellarator WEGA, where the plasma radiation in the visible range shows the confining nested flux surfaces in the center and the magnetic islands at the edge. Reprinted from Ref. [39], with the permission of AIP Publishing.

ICWC was studied on the JET ITER-like wall (beryllium main chamber wall and tungsten divertor) for its applicability to recover hydrogen isotopes [16]. This is of particular interest for ITER. ITER will work with strict in-vessel inventory limits for radioactive tritium. These inventory limits can be reached within a few thousands of discharges, due to codeposition and implantation on the PFCs, if no build-up mitigation efforts are applied [40]. The isotope removal rate by ICWC, averaged over multiple discharges, was shown to be comparable to the upper estimate for fuel retention during fusion plasmas on JET, namely 1.5×10^{20} atoms/s. Moreover, the IC discharge was successfully initiated and sustained over a large range of toroidal magnetic field values: 0.16 T to 3.3 T.

Figures 6.4b and 6.4c show similar tangential camera images for the stellarator magnetic field configuration, taken in WEGA at IPP Greifswald, Germany [39]. The discharge was produced by coupling 2.7 kW of IC power at 9 MHz and a toroidal field of 0.5 T with a field line pitch of $\iota = 0.3$, for a helium neutral pressure of 5.5e-4 mbar. The presence of the confining nested flux surfaces in the center and the magnetic islands at the edge are visible in the plasma radiation pattern. Nevertheless, the confined stellarator discharge, with an electron density on the order of 1e17/m^3, was found as efficient for hydrogen removal as a discharge without a field line pitch ($\iota = 0$) on the same device, that is, with a tokamak field geometry. This is understood by the main role of charge exchange neutrals in conditioning by an RF plasma.

6.3.4 MW Conditioning in Tokamaks

MW conditioning discharges are produced by coupling RF waves at the ECR condition (first harmonic) or its second harmonic at a gas pressure of typically 10^{-5}–10^{-4} mbar. ECR discharge cleaning was first carried out in JFT-2 at the Japan Atomic Energy Research Institute by using the 2.45 GHz LH system and a magnetic field of 87.5 mT. Since then it has been tested on many other devices in studies on reactive and nonreactive cleaning, thin film deposition, and removal of codeposited layers. Superconducting devices such as the W7-X stellarator and the JT-60SA tokamak produce the EC conditioning discharges at their nominal magnetic field and the MW frequency of their EC heating and current drive systems. This section focuses on MW conditioning in tokamaks. These plasmas are substantially different from their counterpart in stellarators due to their strongly reduced plasma confinement. EC conditioning on stellarators is discussed in Section 6.3.5.

Experiments in the TCV tokamak at EPFL in Switzerland assessed the applicability of EC conditioning in helium at the second EC harmonic to desaturate the carbon-based plasma-facing surfaces from deuterium. The work was part of the preparations for the first operations campaign on JT-60SA. Figure 6.5 illustrates a discharge optimization sequence in such an experiment, where small changes to the discharge parameters result in significantly different density profiles, here measured by vertical interferometry lines and plotted as a function of the radial coordinate. The location of the second harmonic EC resonance layer is indicated by the vertical dashed line. The MW power is absorbed by electrons at this layer and transported by the plasma from this position toward the high field side by Bohm-like diffusion and to the low field side by ExB drifts. Applying a small vertical field component in addition to the toroidal field increased the density by 5 times (red vs. blue). Electron currents, which can now flow along the titled magnetic field lines, compensate the build-up of electric fields in the plasma that drive ExB drifts. The increased density due to slower plasma transport, or better plasma confinement, is amplified by an improved EC absorption efficiency at higher densities. Indeed, for extraordinary second harmonic EC waves, the single-pass absorption scales as $n_e T_e$, that is, the product of the electron density and its temperature, at the resonant condition. Changing the angle of the MW beam with respect to the resonance layer also affects the EC absorption efficiency (yellow vs. red). In this discharge the neutral gas becomes fully ionized, marked by an increased plasma temperature that impacts positively the EC absorption efficiency. The discharge becomes moreover volumetric, with nearly equal densities at the low field side and the high field side. Its density maximum is shifted to the center of the vessel at the left side of the resonance layer. When the neutral gas flow into the vessel is finally increased by 2 times, one retrieves again a partially ionized plasma (purple vs. yellow) with a low plasma density at the high field side, which results in poor conditioning results at this location.

The strong nonlinear dependencies on the discharge parameters make predicting the optimal operation conditions for EC conditioning on a new fusion device difficult. Moreover, the best double pass absorption efficiencies in the pulses discussed above (yellow) remained well below 25%. A fraction of the non-absorbed beam power is absorbed on the opposite vessel wall while the majority will be reflected. Care must be taken to avoid absorption of the stray EC energy on sensitive in-vessel components. JT-60SA considers for this reason the operation of EC conditioning at the fundamental first harmonic instead of the second harmonic, with the advantage of higher single-pass absorption efficiencies for the envisaged EC plasma parameters.

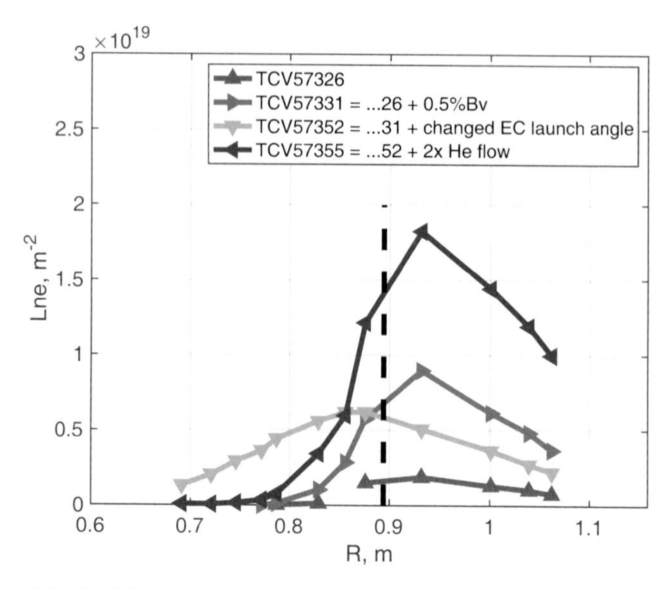

Figure 6.5 Radial density profiles of EC conditioning plasma in TCV measured by vertical interferometry chords, where each subsequent discharge has one parameter changed compared to the previous discharge. Blue: a pure toroidal magnetic field; red: a vertical field component added; yellow: EC launch angle changed; purple: neutral gas throughput increased.

6.3.5 High Temperature Plasma for Conditioning Purposes

The divertor in a thermonuclear fusion reactor serves to exhaust fusion products and impurities. Most current tokamaks and stellarators are operated in divertor configuration. Here the separatrix separates the closed flux surfaces from the region with open field lines, the SOL, as illustrated in Fig. 6.6. The main solid surface that interacts with the plasma, namely the divertor target plates, is located at some distance from the last closed flux surface. Outside the separatrix, the plasma will flow toward the divertor targets. The energy of the flowing particles is partially deposited on the target plates and partially radiated due to the presence of recycled neutrals or seeded impurities. Eventually released impurities from the target will be ionized and swept back with the plasma flow to the target before they can reach the last closed flux surface. This is also the case for material eroded from the main vessel PFC, limiting overall the impurity contamination of the central plasma. The migration of material toward the divertor, and next over the divertor target plates by re-erosion and redeposition, will lead to distinct erosion- and deposition-dominated areas on these plates. Erosion-dominated areas are typically located around the strike lines that endure the high and energetic flux of the main plasma species. In deposition-dominated areas, fuel particles will be codeposited with the impurities to form thick layers rich with hydrogen isotopes. This is of concern (i) for density control, in case these layers are heated in long discharges and start outgassing via thermodesorption, (ii) to the control of the isotopic content of the plasma, and (iii) for the in-vessel inventory of tritium in ITER.

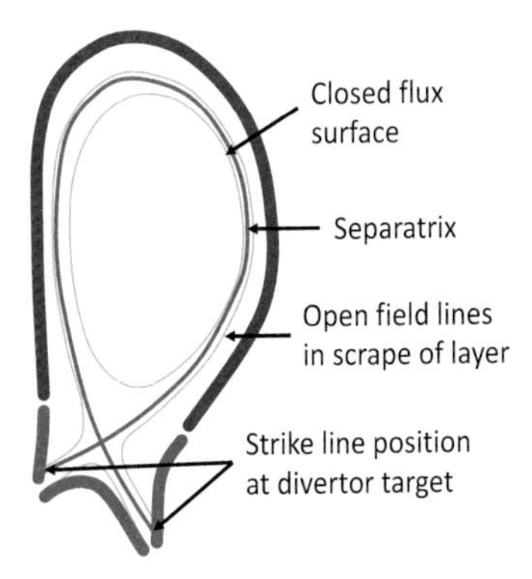

Figure 6.6 Illustration of the divertor configuration. The plasma-facing materials are drawn by the thick lines: dark blue for the main wall and light blue for the divertor target plates. The thin lines map the magnetic field lines on the poloidal cross section.

Scanning the divertor strike line positions over the targets is an effective conditioning method to alleviate these concerns. Such diverted conditioning plasma is distinguished from normal operation by its optimization for removal rather than for stored energy in the plasma core [16]. W7-X uses divertor conditioning by helium plasmas to desaturate the graphite targets from hydrogen and to achieve stable low-density operation with boronized walls. ITER considers the use of divertor conditioning discharges with raised strike points to heat and remove tritium-rich deposits on the divertor targets. Pulsed conditioning by repeated 2 s long diverted discharges with a 10% duty cycle, as routinely performed in W7-X, is however not possible on a tokamak. The repetition rate for tokamak plasma is limited by the cycling of the central solenoid, which drives the plasma current, data collection, and systems controls.

High-temperature plasma is being investigated as well for the application of wall coatings [16]. Boronization through GDC requires de-energized coils and, moreover, strict safety measures regarding the toxic and explosive diborane gas. Dispersing boron particulates in the SOL of diverted plasmas allows one to bring coating material directly to the divertor area and to the poloidal limiters. Beneficial effects of real-time conditioning, namely injecting these powders into long high-performance discharges, is also investigated, where efforts have been made to keep up the plasma purity and minimize the total powder throughput to avoid the accumulation of thick deposits of powder material in long pulses.

6.3.6 Taylor Discharge Cleaning

Tore Supra has long been the only tokamak in the world that operated with a permanent magnetic field provided by superconducting magnetic field coils. To desaturate the carbon-based PFCs from hydrogen during an experimental program and to recover operations after a disruption, they have adopted the Taylor discharge cleaning (TDC) technique. TDC, first developed by Taylor [17], relies on weakly ionized plasma produced by pulsing the central solenoid of a tokamak. The plasma current is limited to a value below ~30 kA to maintain partially ionized low-electron-temperature discharges.

TDC will not be used on the future fully superconducting device ITER since the superconducting central solenoid is not designed to withstand the pulsed voltage operation required for TDC. The risks involve heating up the conductors, which may lead to the loss of their superconductivity. Secondly, the high loop–voltage consumption would require repetitive charging of the central solenoid, which would take typically 100 s of charging for creating a 1 s TDC pulse.

A comparison on TORE SUPRA between TDC and ICWC in helium for desaturating the carbon-based PFCs showed that the hydrogen removal efficiency of ICWC is 2 to 5 times higher than that of TDC discharges. Since TDC is successfully used for interpulse conditioning on TORE SUPRA, this result indicates that He-ICWC can also be used as its replacement for interpulse conditioning on future devices.

6.4 Plasma Coating

Fast and efficient modification of the PFCs in fusion devices can be achieved by coating the inner elements with suitable materials. Although in many instances this can be done by in vacuo evaporation, as is the case with Be, Ti, and Li getters [13–15], a more homogenous and controlled distribution is typically obtained by using plasma-asisted chemical vapor deposition techniques, as those described in previous chapters for the microelectronic industry. Carbon, boron, and silicon films have been historically applied to the first wall of several fusion devices with different degrees of success [18]. For a given coating species, different molecular precursors can be injected into the main conditioning plasma, resulting in films with different properties. This has been extensively studied for boron coatings [19]. Some of the film characteristics of relevance for plasma operation are its chemical composition in terms of H/B and C/B ratios, ability to stick to the wall, erosion yields, chemical reactivity with vacuum background gases, and porosity. The background plasma has also a big impact on the achieved properties, including its nature (He, D_2, Ar) and type (see above). This point is of particular relevance when dealing with the implementation of these techniques for superconducting devices. Table 6.2 summarizes some important findings obtained in different machines using MW or RF plasmas for boronization of the first wall.

In spite of the success achieved by the deposition techniques using cold plasmas for low Z_{eff} coating in fusion devices, there are still some problems to solve. On one hand, a homogeneous film coating is difficult to obtain in devices with complicated geometry or large vacuum vessels. To improve that, it is important to have a higher number of gas entries and to work in conditions of partial cracking so that the residence time of the precursor is long enough for it to spread uniformly through the full vessel. This subject is addressed in some more detail in the practical example given in Section 6.5.

Table 6.2 Some examples of B deposition in the presence of a magnetic field

Device	Plasma	Precursor	Other	Findings
EAST	RF 30 MHz, 20 KW	Carborane (+He)	T_{wall} = 180°C, B = 1–2 T	Lifetime coating (300 nm) ~400 shots
TEXTOR	DC + RF, 29 MHz, 100 W	Diborane (+He)	B = 2.5 T, inhomogeneous	Lifetime (100–200 nm) = 100s shots
Heliotron-E	ECD, 2.45 GHz, 2 kW	Decaborane, diborane (+He)	RT, 3 inlets, B = 0.045 T	Decaborane → faster deposition but high H/B
Alcator C	ECD, 2.45 GHz, 3 kW	Diborane (+He)	B = 0.1 T, distributed source	ECD films with less H content than DC GD

On the other hand, since it is not possible to constrain the coating to a particular area of the vessel, the same kind of film is produced elsewhere. However, the plasma-wetted areas, as the limiter or divertor target surfaces, are exposed to far more erosion than those in the main vessel walls, only reached by charge exchange neutrals and photons. One of the possibilities to localize the deposition on the heavily eroded areas and to avoid discontinuity in the regular device operation concomitant to the film deposition is to use real-time wall conditioning techniques. The injected film precursor is fully ionized into the plasma edge and then transported to the areas of stronger interaction. In the case of lithium coatings, a few ideas have been proposed and successfully tested [20]. However, for real-time boronization, only the interaction of solid boron with hot plasma has been attempted, with limited success [21]. Very recently, the same technique used for lithium coating, based on powder dropping during hot plasma operation from a piezo-electrically driven valve, has been implemented for real-time boronization in the ASDEX Upgrade tokamak [22]. Promising results were obtained, calling for a devoted effort aimed at the understanding and improvement of this reactor-compatible conditioning technique.

6.5 A Practical Case: Wall Conditioning of the TJ-II Stellarator

The TJ-II stellarator is being operated since 1997 in the Laboratorio Nacional de Fusión-CIEMAT, in Madrid [23]. Contrary to tokamaks, stellarators use external coils without a plasma current to produce all the magnetic fields necessary for the plasma confinement. As a consequence of that, stellarators present complicated geometries that make the conditioning of the vacuum chamber rather challenging.

TJ-II is considered a medium size device with a major radius $R = 1.5$ m and an average minor radius (plasma size) $a \leq 0.22$ m. The magnetic field $\mathbf{B}_0 \leq 1$ T. Plasmas are produced and heated by electron cyclotron resonance heating (ECRH) with power up to 600 KW in two lines working at 53.2 GHz. Additionally, two lines of neutral beam injectors (NBIs) are operated with power up to 1 MW.

The vacuum vessel, all stainless steel, has a volume of 6 m^3, and the total inner area is 75 m^2. Figure 6.7 shows a sector of the vacuum vessel and a poloidal view of the plasma-wall interaction area. The interaction of the plasma with the chamber is localized to the groove surrounding the helical coils that acts like a helical limiter.

The close coupling of the vacuum vessel and the plasma periphery combined with the small volume of TJ-II plasmas (less than 1.4 m^3) with respect to the large vacuum vessel surface area results in desorbed impurities during the discharge, making an important contribution to the plasma density. This strong interaction relatively near the plasma center and the low value for the ECRH cut-off density limit (10^{13}cm^{-3}) have conditioned the operational limits of TJ-II since the beginning of the experiments [24]. Therefore, the procedures of wall conditioning have been critical in the operation of the device.

Baking of the vacuum vessel (T up to 150°C) was performed before the 1999 campaign, with good results in the base pressure improvement. However, due to the technical difficulty of the procedure with an increasing number of peripheral systems in the machine, it was recommended that the vacuum chamber not be baked again. Therefore, all the wall conditioning procedures described below were performed at room temperature.

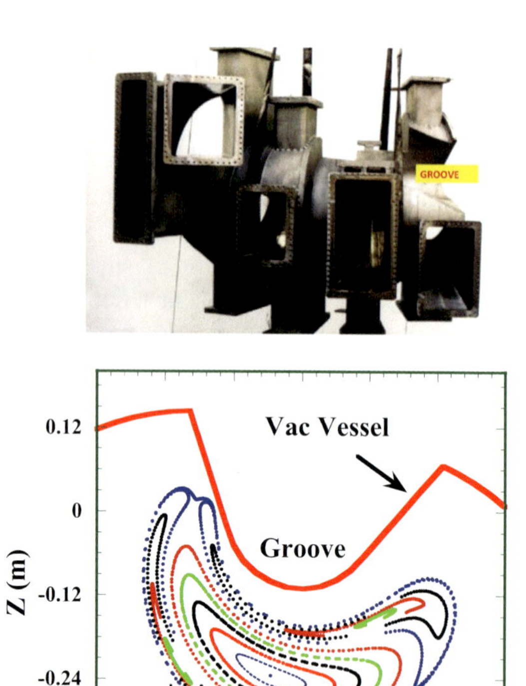

Figure 6.7 (Top) sector of the TJ-II vacuum vessel showing the indentation (groove) coupling of the vacuum vessel to the central coil system. (Bottom) poloidal view of the plasma (magnetic surfaces) showing the strong interaction between the plasma and the vacuum vessel in the groove region.

6.5.1 He Glow Discharge during the First Campaigns of TJ-II

During the initial operation campaigns (1997–2001) the TJ-II stellarator was operated with stainless-steel plasma-facing surfaces, and the discharges were heated only by ECRH. The device is equipped with normal conducting coils that are energized for the duration of the plasma discharge. One of the first problems was the generation of fast electrons during the initial ramp-up of the currents in the coil system. Contrary to the tokamak case, the presence of nested magnetic surfaces during the plasma initiation in stellarators led to the generation of fast electrons. The runaway electrons produce hard X-ray emission before the ECRH discharge, affecting the plasma reproducibility and the normal behavior of different diagnostics. Their production correlates directly with the residual pressure of the vacuum chamber, becoming systematic for pressures above 4×10^{-7} mbar. Also, the presence of high values of hydrogenic species (e.g., H_2O), in comparison with others, like He, increases the production of fast electrons.

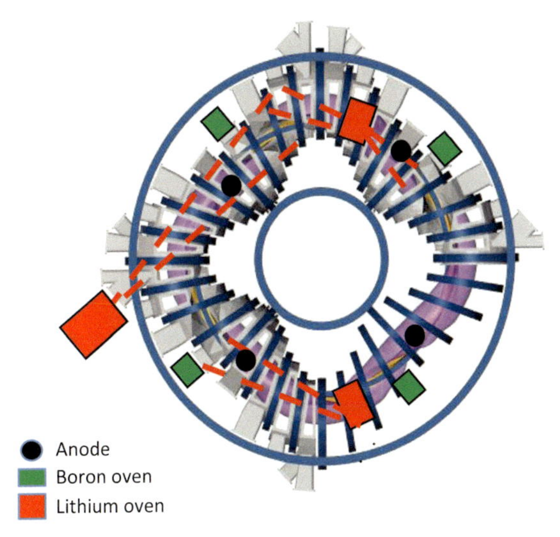

Figure 6.8 Schematic view of the TJ-II stellarator indicating the positions of the anodes, the boron ovens, and the lithium ovens used for the wall conditioning procedures.

The wall conditioning procedure in this period was an intensive helium GDC during overnight periods (about 14 h) [25, 26]. The GD was carried out by applying a DC voltage of about 300 V to two (extended to four in 2001) L-shaped stainless-steel anodes, located toroidally distributed at the low field side of the vessel (see Fig. 6.8 for the positions). The total discharge current was typically 2 A, equivalent to a current density of about 2.5 mA cm^{-2}. The discharge was started up by applying a DC voltage of 1000 V and with the help of a short injection of argon. When the glow was started, the argon injection was switched off and the discharge sustained at helium pressures in the range of 3×10^{-3} to 5×10^{-3} mbar using a feedback-controlled gas injection system. At this pressure, the pumping speed of the vacuum system was about 2400 l/s. A residual gas analyzer (RGA) of the quadrupole type was used to analyze the evolution of gas species produced during the conditioning discharge.

The main effects of overnight He GDC under these conditions were (i) removal of the hydrogen implanted on the walls by the plasma discharges of the previous experimental day and (ii) production of an activated surface with a typical wall pumping behavior. The last effect is observed by the decrease in the residual pressure by about 10%–20% after the He GDC. The recovery time of water and oxygen after the He GDC was found to be dependent on the number of previous GDC cycles and the TJ-II shots, increasing as the experimental campaign progressed. The decrease in the TJ-II base pressure by He GDC had a critical effect on the TJ-II operation. The strong depletion of hydrogenic species in the residual vacuum resulted in the suppression of fast electrons during the ramp up of the currents in the coils and the control of X-ray emission before the ECRH discharge. As such, good control of the start-up of the discharge was reached and reproducible plasma discharges were obtained.

However, this procedure gave rise to a lack in density control when the ECRH-injected power was higher than 300 kW. Under these conditions, the He implanted on the walls during GDC is desorbed by plasma discharges, producing uncontrolled increase of the plasma density, which reaches the cut-off limit. To remove the implanted He, we started to apply a short Ar GDC (<30 min) after the overnight He GDC. The release of He atoms from the walls by Ar bombardment is shown in Fig. 6.9. The Ar bombardment allowed the release of about 10^{21} He atoms, equivalent to 10^{15} atoms/cm^2, considering the total surface of the TJ-II vacuum chamber. The He depletion so obtained allows good density control by external gas puffing under high injected ECRH power [26].

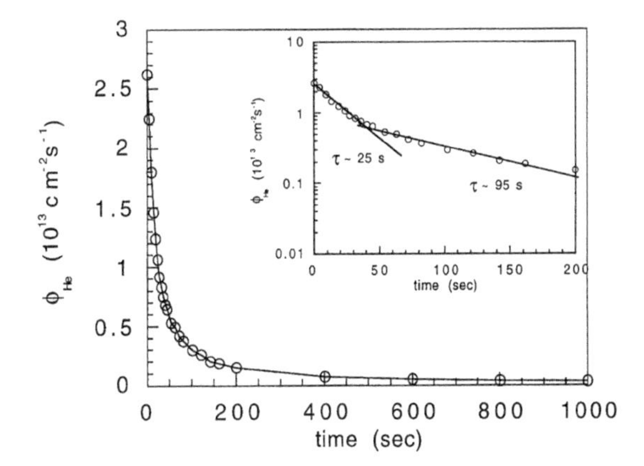

Figure 6.9 Released He flux density during the Ar GDC subsequent to the overnight He GDC. The inset shows the initial decay of the signal on a semilogarithmic scale. From these data, a value of the detrapping cross section could be estimated. Reprinted from Ref. [25], Copyright (2002), with permission from Elsevier.

6.5.2 Boronization of TJ-II

The starting of NBI operation in TJ-II in 2003 required a change in the wall conditioning procedure. In this new heating scenario, the particles and thermal fluxes to the different parts of the vacuum vessel increases and a significant enhancement of the metallic sputtering is produced. The use of graphite poloidal limiters and graphite thermal protections required to absorb the fraction of neutrals, and their energy, that may travel freely through the plasma, precludes the use of extensive GDC in TJ-II.

Plasma-assisted deposition of amorphous boron or carbon-boron hydrogenated thin films (a-B/C:H), a process called boronization, has been extensively used in fusion devices during the last two decades [7]. In addition to the suppression of metallic impurities, the main advantage of B/C films over pure carbon ones is the ability to getter oxygen by boron and their superior resistance to chemical erosion. Both effects produce a significant reduction in the carbon and oxygen content in the plasma.

The initial method of boronization, and probably the most practiced in large fusion devices, is by means of GD using admixtures of B_2H_6, an explosive and highly toxic gas that requires special equipment and operation procedures to ensure safe handling. The use of less dangerous substances has been extended in the last years, as an alternative to diborane, especially in small and medium size devices. Among these substances, the *o*-carborane ($C_2B_{10}H_{12}$), a nontoxic and non-explosive solid (powder) with a high vapor pressure, seems to be the most promising for simple and low-cost boronization [19]. In our laboratory, another two substances (trimethyl-boron and decaborane) were tested for boronization in a previous device [27], but finally *o*-carborane was chosen because of its ease of use. In summary, the boronization procedure in TJ-II consisted of injecting an amount of *o*-carborane during a He GD [28].

Challenging for a device with a complicated geometry like TJ-II is obtaining uniform B/C films. For this purpose, four ovens were installed in symmetrical positions in the TJ-II vacuum vessel (see Fig. 6.8 for the positions). A total of 4 g of carborane (1 g per oven) was injected during He GD. This amount is enough to cover the inner wall exposed to plasma with a B/C film of 50 nm in the case of a totally homogeneous deposition. Carborane injection was produced by heating the ovens continuously to a temperature of 100°C. The process was monitored by an RGA. The continuous ramp-up of the temperature made it difficult to estimate the cracking efficiency from the RGA data during the deposition. The presence of mass 70 and 144 in the mass spectra indicated that the molecule was not completely cracked, as required in order to obtain a good uniformity in the deposition film. The cracking of the precursor (N/N_0) can be expressed by the simple equation

$$N/N_0 = 1/(1 + K_d n_e \tau_r). \tag{6.3}$$

As can be seen, the microscopic plasma parameters n_e and T_e, the dissociation constant of the molecule K_d (dependent on T_e, and the residence time of the vacuum system $\tau_r = V/S_{eff}$ determine the cracking factor. The cracking factor has to be lower than 100% so a homogeneous coating can be obtained. In TJ-II a value of 80% was estimated, indicating a fairly homogeneous coating. No condensation of carborane on the walls or the injection line was observed after deposition. However, the apertures of the TJ-II vacuum vessel in 2017 (after about 70 boronization processes) showed preferential deposition in the sectors near the boron inlet, indicating that not all the depositions were highly homogeneous. We are currently trying to improve this homogeneity by mainly controlling the cracking during the deposition by means of the plasma current and measuring the deposited boron using laser-induced breakdown spectroscopy techniques [29].

The boronization is always followed by a period of 30 min of He GDC in order to remove a part of the hydrogen retained in the film. The B/C film has an immediate gettering effect on the oxygen-containing molecules (O_2, H_2O, CO, CO_2), and these species almost disappear from the residual gas spectrum. The effect diminishes after one day of ECRH plasmas, but it can be regenerated with a 30 min. He GD every day for several weeks (even several months, depending on the plasma discharges). No evidence of film erosion, in the form of B compounds released, was found in pure He GD [28].

As a result of the gettering effect, there was a strong reduction in light impurities (C, O) in the TJ-II ECRH plasmas and a concomitant decrease in the radiated power as measured by bolometers and soft X-ray monitors. Besides, the high capacity of the film for hydrogen retention produced a strong variation in the recycling behavior. It was necessary to inject a double amount of H_2 in order to obtain a plasma density equivalent to the all-metal case. Although the density control presented a high dependence on the wall condition, a good control of the discharge was reached in a few shots. Especially remarkable was the effect on the plasma edge temperature, which reached values nearly twice the those of previous ones in TJ-II [30].

Although the operation of ECRH plasmas was highly enhanced by the B coating, the plasma operation at "high" density (near the cut-off), especially in the NBI heating phase, was hindered by poor control of recycling. This is a consequence of the rapid saturation of the film by hydrogen surface densities on the order of 10^{17} cm^{-2} [31]. Only by dry discharges (no "puffing") or the application of He GD it was possible to recover the initial good recycling condition. This is shown in Fig. 6.10, where the total particle inventory (hydrogen atoms) is represented. A limit of wall saturation of around 4×10^{20} particles is observed. When this value is reached (with about 20 plasma shots) the density control is ineffective and transitions to plasma densities about the cut-off limit are systematically produced.

He GDC applied for 30 min before the shot #12300 in data of Fig. 6.10 produced a return to good recycling conditions, and a saturated wall is attained again after about 20 shots.

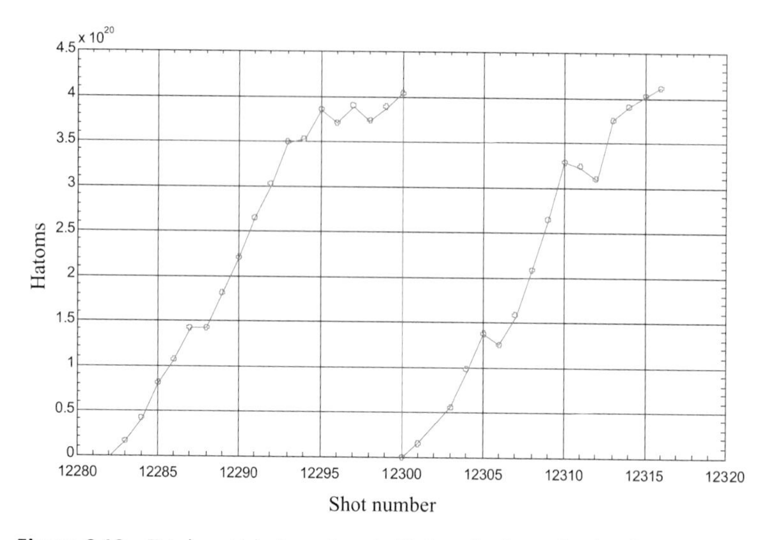

Figure 6.10 Total particle inventory in TJ-II under boronized walls. Reprinted from Ref. [35], Copyright (2010), with permission from Elsevier.

Due to these problems and the physics requirement of increasing the injected power by NBI, the possibility of applying a coating with active hydrogen pumping was considered. The use of lithium seemed the best option because of the low Z and the reactivity of the metal. Although lithium had been used in several tokamaks with some success [32], it had not been applied so far in stellarators.

6.5.3 Lithium Coating in TJ-II

The lithium coating of the TJ-II vacuum vessel was tested in the experimental campaign of 2007–2008 [33] and has been widely applied since. For that purpose, an in situ lithium coating technique at room temperature was developed (lithiation). It is based on the evaporation of metallic Li under vacuum from three retractable ovens, symmetrically spaced and oriented tangentially to the vacuum vessel in the equatorial plane of the machine. In the Fig. 6.8 the position of the oven and the deposition direction are indicated.

A fourth oven was finally removed again to avoid the interaction of the NBI beam with the fresh lithium coating [34]. A total of 3 g of metallic Li (1 g per oven) is evaporated during each conditioning cycle, at temperatures of 500°C–600°C. Effusion from the ovens creates an atomic beam aiming at the remote region opposed to the corresponding flange. Under high vacuum operation, the mean free path of the Li atoms is long enough to produce a thin layer at the vessel walls located midway between adjacent ovens. The initial deposition pattern, directly visible in the groove protecting the central coils, matches the free trajectory of the Li atoms. Attempts to enhance the spreading of Li by increasing the background pressure didn´t result in a better performance. Moreover, the addition of a Ne GD plasma to promote the ionization and redistribution of the evaporated lithium was also investigated. Figure 6.11 shows the oven and the corresponding Li plume under such conditions. Since the pressure required for the Ne GD was rather high, we found local deposition patterns near the ovens. This plasma-enhanced evaporation technique has been successfully used in other devices, but with much lower pressure (MW discharges).

Figure 6.11 Effusion of Li from an oven under a Ne GD plasma at the beginning of the 2009 autumn campaign.

Nevertheless, plasma operation was found to redistribute the initial coating very efficiently and the beneficial effect of the coating extended far beyond the expectations of the localized deposition.

To extend the lifetime of the Li coating, and due to the very high reactivity of this species with background gases (O_2, N_2, CO, etc.), the Li was deposited on the walls previously coated with a B/C film of ~50 nm obtained by the usual boronization procedure described above.

Also, He GDC (<30 min) was applied every day on the Li layer for removing hydrogen from the areas not fully covered by the coating and activating the getter effect of the B film. Figure 6.12 is an example of the changes in the residual gases after Li coating at the beginning of an experimental campaign. The dramatic reduction in the O_2 and H_2O traces evidences the strong gettering effect of the Li coating.

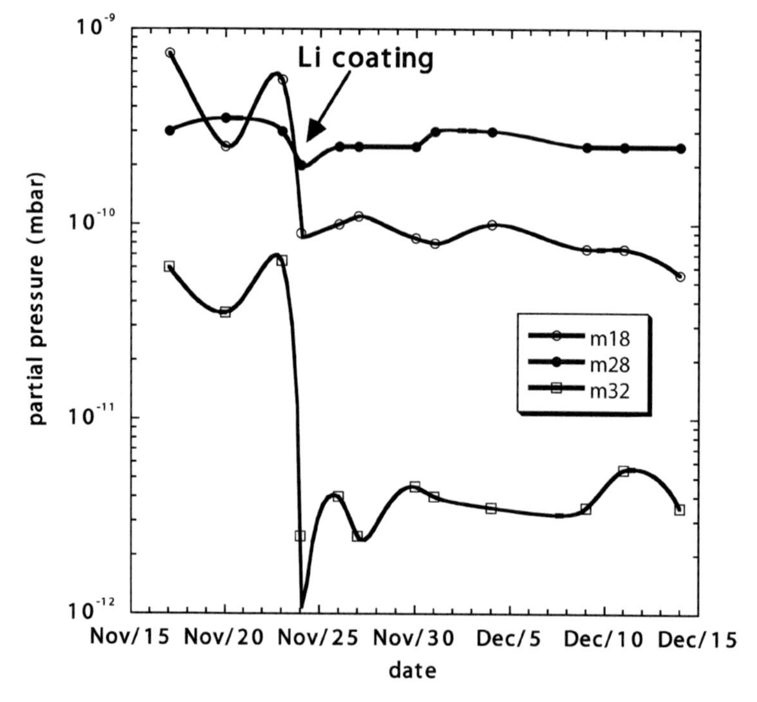

Figure 6.12 Strong decrease in O_2 and H_2O residual pressure after lithiation at the beginning of the 2009 autumn campaign. Reprinted from Ref. [35], Copyright (2010), with permission from Elsevier.

With less O-containing molecules in the neutral gas follows also a reduction in the impurity generation during plasma discharges.

The radiated power measured by bolometers shows a strong decrease the first operational days after lithiation, and this decrease increases during the plasma operation [35], as can be seen in Fig. 6.13. This effect could be explained by the erosion of the Li in the initial layer accessible through the observation window (note the decay of normalized lithium emission) with subsequent spreading over other areas of the vessel, in agreement with visual inspection of the initially wetted area.

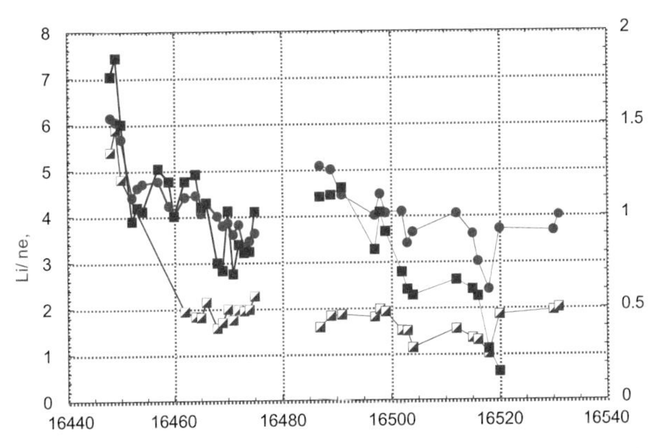

Figure 6.13 Impurity evolution during the first two operation days after lithium deposition. A He GD was used for film activation before each day of operation. Reprinted from Ref. [35], Copyright (2010), with permission from Elsevier.

After lithiation a strong improvement in particle control by external gas injection was observed. The required puffing levels were higher than in the B-coated case by a factor of 2 to 3 to obtain a similar density. Also, with Li-coated walls, no sign of wall saturation was observed after a full day of ECRH plasmas, as shown in Fig. 6.14. Particle balance measurements yields a total retention of $\sim 4 \times 10^{21}$ H atoms, 6 times higher than with B coating.

In general, the TJ-II operation after Li coating shows an excellent density control by external gas puffing and a high reproducibility of the plasma discharges. Because of Li coating, TJ-II has achieved, for the first time, up to 900 kW of NBI-injected power and shots with electronic density of up to 5×10^{19} m^{-3} with a stored energy of 5 kJ.

The extension of the operational window introduced new plasma features that have not been obtained in TJ-II so far, most importantly the L-H mode transition [36].

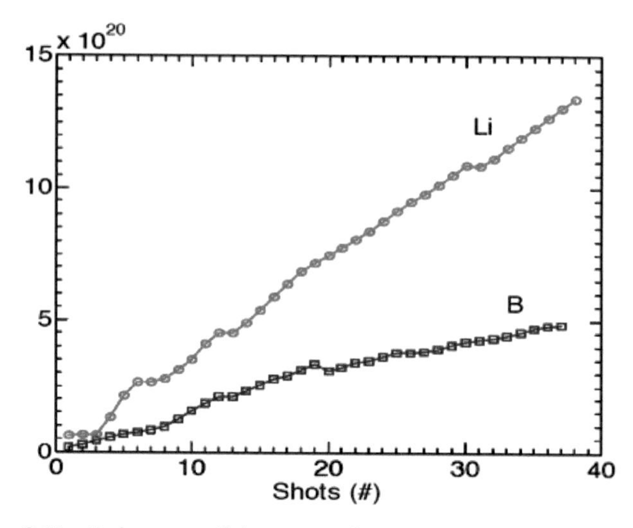

Figure 6.14 Hydrogen wall inventory during an operational day for the B-coated and the Li-coated wall. Reprinted from Ref. [35], Copyright (2010), with permission from Elsevier.

6.6 The Future: Wall Condition of Reactor-Oriented Fusion Devices

Although the peculiarities of wall conditioning plasmas of use in a superconducting device have been commented on all through the previous sections, their relevance for reactor-oriented machines deserve a brief summary. A thorough description and analysis of the topic can be found in Refs. [8] and [16]. Also, Table 6.1 gives relevant information in this respect. In brief, the constraint imposed by the superconducting coils makes DC GD–based methods for wall conditioning and coating almost irrelevant. Only during long stops and in the absence of the superconductor (SC) field may these techniques have some use. Otherwise the cost and side effects implied in the de-energizing of the coils will not be justified. Nevertheless, in the very early phase of machine start-up, baking and long DC GD plasma conditioning may become mandatory.

Routine operation of the SC plasma device calls for isotope and impurity control, as demonstrated in present-day fusion devices. This can be accomplished by the EM-wave-based techniques mentioned in Section 6.2. Present knowledge point to ICWC as more flexible than the electron cyclotron wall conditioning counterpart, as resonance absorption is not required for ICRF waves. A different situation may apply, however, for plasma breakdown assistance, where similar EM plasmas are used to preionise the neutral gas before the plasma current is started. Here a larger database exists for ECRH at present.

The divertor region concentrates most of the exhaust of particles and power of the plasma. One of the main concerns in ITER, with a Be first wall, is the removal of T trapped in codeposits with the eroded Be at the divertor area. After a score of experiments devoted to the topic, it seems that the most effective technique for the removal of these codeposits is baking the divertor at $T > 350°C$. However, its efficiency drastically drops as the thickness of the deposits increases, and unpractically long baking times would be required for codeposits of over 50 μm [37]. In addition, the presence of W or other impurities on the codeposits, or their exposure at temperatures near 350°C, may jeopardize the success of this simple removal method. Active research on the topic, including the ability of conditioning plasmas for selective isotopic removal/exchange, is underway.

Acknowledgments

This work has been partially carried out within the framework of the EUROfusion Consortium and has received funding from the Euratom Research and Training Programme 2014–2018 and 2019–2020 under Grant Agreement No. 633053. The views and opinions expressed herein do not necessarily reflect those of the European Commission.

References

1. D. E. Post, R. Behrisch (eds.), *Physics of Plasma-Wall Interactions in Controlled Fusion*, in NATO ASI Series B: Physics, Plenum Press, New York (1986).

2. International Atomic Energy Agency, Atomic and plasma-material interaction data for fusion, Suppl. to *Nucl. Fusion* (several issues).
3. G. Federici et al., Plasma-materials interactions in current tokamaks and their implications for next steps fusion reactors, *Nucl. Fusion*, **41** (2001) 1967.
4. C. S. Pitcher, P. C. Stangeby, Experimental divertor physics, *Plasma Phys. Control. Fusion*, **39** (1997) 779.
5. P. C. Stangeby, *The Plasma Boundary of Magnetic Fusion Devices*, Institute of Physics Publishing, Bristol, England (2000) p. 712.
6. H. F. Dylla, Glow discharge techniques for conditioning high-vacuum systems, *J. Vac. Sci. Technol. A*, **6** (1988) 1276.
7. J. Winter, Wall conditioning in fusion devices and its influence on plasma performance, *Plasma Phys. Control. Fusion*, **38** (1996) 1503.
8. D. Douai et al., Wall conditioning for ITER: Current experimental and modeling activities, *J. Nucl. Mater.*, **463** (2105) 150–156.
9. D. Reiter, P. Bogen, U. Samm, Measurement and Monte-Carlo computations of H_2 profiles in front of a TEXTOR limiter, *J. Nucl. Mater.*, **196–198** (1992) 1059.
10. A. Hassanein, Disruption damage to plasma-facing components from various plasma instabilities, *Fusion Technol.*, **30**(3P2A) (1996) 713–719.
11. D. Kogut et al., Modelling of tokamak glow discharge cleaning II: comparison with experiment and application to ITER, *Plasma Phys. Control. Fusion*, **57** (2014) 025009.
12. T. Härtl et al., Optimization of the ASDEX upgrade glow discharge, *Fusion Eng. Des.*, **124** (2017) 283–286.
13. J. Ehrenberg, V. Philipps, L. DeKock et al. Analysis of deuterium recycling in JET under berillium first wall conditions, *J. Nucl. Mater.*, **176–177** (1990) 226.
14. A. J. Wooton, P. H. Edmonds, R. C. Isler, P. Mioduszewski, Gettering in ISX-B, *J. Nucl. Mater.*, **111–112** (1982) 479.
15. F. L. Tabarés et al., Plasma performance and confinement in the TJ-IIstellarator with lithium-coated walls, *Plasma Phys. Control. Fusion*, **50** (2008) 124051.
16. T. Wauters et al., Wall conditioning in fusion devices with superconducting coils, *Plasma Phys. Control. Fusion*, **62** (2020) 034002.
17. L. Oren., R. Taylor, Trapping and removal of oxygen in tokamaks, *Nucl. Fusion*, **17** (1977) 1143.

18. J. Winter et al., Boronization in textor, *J. Nucl. Mater.*, **162–164** (1989) 713.
19. O. I. Buzhinskij, Y. M. Semenets, Review of in situ boronization on contemporary tokamaks, *Fusion Technol.*, **32** (1997) 1.
20. D. K. Mansfield et al., A simple apparatus for the injection of lithium aerosol into the scrape-off layer on fusion research devices, *Fusion Eng. Des.*, **85** (2010) 890.
21. C. Boucher et al., Comparison of three boronization techniques in TdeV, *J. Nucl. Mater.*, **196–198** (1992) 587–591.
22. A. Bortolon et al., Real-time wall conditioning by controlled injection of boron and boron nitride powder in full tungsten wall ASDEX upgrade, *Nucl. Mater. Energy*, **19** (2019) 384–389.
23. C. Alejaldre et al. First plasmas in the TJ-II flexible Heliac, *Plasma Phys. Control. Fusion*, **41** (1999) A539.
24. F. L. Tabarés, D. Tafalla, E. de la Cal, B. Brañas, Plasma-wall interactions in the Spanish stellarator TJ-II. Diagnostics and first results, *J. Nucl. Mater.*, **266–269** (1999) 1273.
25. D. Tafalla, F. L. Tabarés, Wall conditioning of the TJ-II stellarator by glow discharge, *Vacuum*, **64** (2002) 411.
26. D. Tafalla, F. L. Tabarés, Wall conditioning and density control in theTJ-II stellarator, *J. Nucl. Mater.*, **290–293** (2001) 1195.
27. F. L. Tabarés, E. de la Cal, D. Tafalla, Boronization of the TJ-I tokamak by trimethylboron and decaborane, *J. Nucl. Mater.*, **220–222** (1995) 688.
28. D. Tafalla, F. L. Tabarés, First boronization of the TJ-II stellarator, *Vacuum*, **67** (2002) 393.
29. B. López-Miranda et al., A LIBS method for simultaneous monitoring of the impurities and the hydrogenic composition present in the wall of the TJ-II stellarator, *Rev. Sci. Instrum.*, **87** (2016) 11D811.
30. F. L. Tabarés et al., Impact of wall conditioning and gas fuelling on the enhanced confinement modes in TJ-II, *J. Nucl. Mater.*, **313–316** (2003) 839.
31. J. Ferreira, F. L. Tabarés, D. Tafalla, Particle balance in TJ-II plasmas under boronized wall conditions, *32nd EPS Conference*, Tarragona, Spain (2005).
32. S. Mirnov, Plasma wall interactions and plasma behaviour in fusion devices with liquid lithium plasma facing components, *J. Nucl. Mater.*, **390–391** (2009) 876.

33. J. Sánchez et al., Impact of lithium-coated walls on plasma performance in the TJ-II stellarator, *J. Nucl. Mater.*, **390–391** (2009) 852.

34. M. Liniers et al., Beamline duct monitoring of the TJ-II neutral beam injectors, *Fusion Eng. Des.*, **88** (2013) 960.

35. D. Tafalla, F. L.Tabarés, J. Ferreira, Wall conditioning strategies in the stellarator TJ-II, *Fusion Eng. Des.*, **85** (2010) 915.

36. T. Estrada et al., Sheared flows and transition to improved confinement regime in the TJ-II stellarator, *Plasma Phys. Control. Fusion*, **51** (2008) 124015.

37. G. de Temmermann et al., Efficiency of thermal outgassing for tritium retention measurement and removal in ITER, *Nucl. Mater. Energy*, **12** (2017) 267–272.

38. D. Douai et al., *J. Nucl. Mater.*, Recent results on Ion Cyclotron Wall Conditioning in mid and large size tokamaks, **415** (2011) S1021–S1028.

39. T. Wauters et al., Ion and electron cyclotron wall conditioning in stellarator and tokamak magnetic field configuration on WEGA, *AIP Conf. Proc.*, **1580**(187) (2014); doi: 10.1063/1.4864519.

40. J. Roth et al. Tritium inventory in ITER plasma-facing materials and tritium removal procedures, *Plasma Phys. Controlled Fusion*, **50**(10) (2008) 103001. https://doi.org/10.1088/0741-3335/50/10/103001

Chapter 7

Cold Atmospheric Pressure Plasma Jets and Their Applications

Gheorghe Dinescu and Maximilian Teodorescu
National Institute for Laser, Plasma and Radiation Physics,
Atomistilor Street 409, Magurele Bucharest, Ilfov, Romania
dinescug@infim.ro

This chapter focuses on cold atmospheric pressure plasma sources and their principles of operation, designs, and practical models. Processing examples of surface modification, surface cleaning, thin film deposition, decomposition of chemical substances, and nanomaterial processing by plasma in liquids emphasize their applicative potential.

7.1 Introduction

7.1.1 Meaning of Cold Plasma

The technological plasmas are generally based on electrical discharges in gases. A typical circuit used to generate an electrical

Plasma Applications for Material Modification: From Microelectronics to Biological Materials
Edited by Francisco L. Tabarés

ISBN 978-981-4877-35-0 (Hardcover), 978-1-003-11920-3 (eBook)
www.jennystanford.com

discharge is shown in Fig. 7.1. In consists of a direct current (DC) power supply, a glass tube filled with gas at low pressure (pressure p ~1 mbar), two planar electrodes (cathode C and anode A) separated by the interelectrode distance d (in the range of a few to tens of centimeters), and a series resistor R, named ballast. In normal conditions the gas between the electrodes is insulator, but on increasing the voltage, it suddenly passes into a conducting state, a phenomenon known as electrical gas breakdown. At breakdown the gas is ionized and free electric charges, pairs of electrons and ions, are produced. After breakdown the internal resistance of the gas becomes small and the current in the circuit is practically limited by the ballast resistor (or, in the case of power supplies with current control, by the current set value). In these conditions a plasma is produced in the interelectrode space, consisting roughly of a mixture of charged particles and neutrals, with a quasi-equal number of electrons and ions.

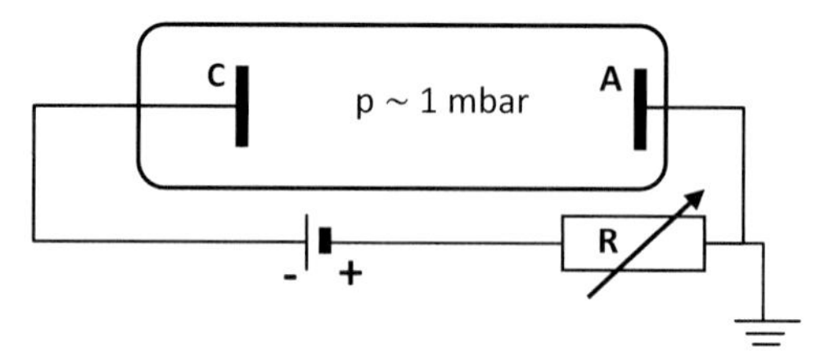

Figure 7.1 A typical electrical circuit used to generate an electrical discharge.

Electron, ion, and neutral gas populations are statistical ensembles of different particles and in a simplified manner can be regarded as three subsystems of a plasma with different associated temperatures T_e, T_i, and T_n. Due to the particularities related to the energy gain of particles from the electrical field (electrons and ions gain energy in the field but neutrals do not) and from the different rates of kinetic energy transfer during collisions (light electrons preserve most of their kinetic energy even after frequent elastic collisions with neutrals, while ions transfer almost half of their kinetic energy in elastic collisions with neutrals), the values of the temperatures associated with the three subsystems of particles are different ($T_e \neq T_i \neq T_n$). For this reason, it is said that a plasma is a non-equilibrium system. In practice, $T_i \approx T_n$ because of the efficient transfer of energy between species with almost equal mass, but $T_e \gg T_i$, T_n. This situation occurs in the majority of plasmas created at a low pressure. Still, there are physical conditions when the values of all three parameters are equal or almost equal, for example, in the case of plasmas created by arc discharges. In those cases, one speaks of equilibrium or thermal (meaning in thermal equilibrium) plasmas, because the physical system is characterized by the same, one temperature:

$$T = T_e \approx T_i \approx T_n$$

The temperature of electrons, which is also named "plasma temperature," is one of the most important parameters of discharge plasmas. Due to the inefficient transfer of their energy by elastic collisions with the heavy particles, the kinetic energy of electrons increases in the electric field until the threshold for an inelastic collision is reached; then, the kinetic energy is transferred to the neutrals by excitation or ionization events. A sufficient number of new ionization events is essential to compensate the charge losses from the discharge volume, and this is a reason why, for maintaining the discharge, the electron temperatures cannot be arbitrarily small. Typical electron temperatures in electrical discharges are about 1–4 eV.

While the electron temperature T_e gives a measure of the system capability to host ionization and excitation events, the neutral (gas) temperature T_n gives a measure of its heat content and its capability to transfer heat to the surroundings. Plasmas with gas temperatures much lower than T_e are named "low-temperature plasmas." In particular, plasmas having a gas temperature not far from ambient temperature are named using a more recent terminology: cold plasmas. Typical values for gas temperatures in cold plasmas are around 0.03 eV.

7.1.2 Importance of Voltage Frequency, Gas Pressure, and Discharge Electrode Configuration in Plasma Sources Operation

Historically, the most studied plasmas were those generated by DC discharges. The development of electric power supplies able to provide alternative or pulsed voltages for various applications (energy transport, communication [radio] technology, radars, etc.) impacted the plasma field, and currently plasmas are also generated by such voltages. Therefore, beside DC discharges, a wide variety of discharges are utilized nowadays: (i) low-frequency discharges (a few hertz up to kilohertz), (ii) high-frequency discharges (kilohertz to hundreds of kilohertz); (iii) radio frequency (RF) discharges (in the megahertz range, typically tens of megahertz [radio waves]), (iv) microwave discharges (gigahertz range; compare this with the microwave oven—2.45 GHz), (v) pulsed electrical discharges (negative or positive pulses with duration from microseconds to nanoseconds) and various repetition rates, and (vi) laser-generated discharges obtained by focalized pulsed plasma beams, with optical breakdown at 10^{14} Hz (these are pulsed plasmas, with pulse durations in the nanosecond or picosecond (using nanosecond and picosecond lasers).

The phenomena in plasmas sustained by alternative (and pulsed) voltages are very special. During change in polarity the anode and cathode in a tube configuration like that presented in Fig. 7.1 change their roles at each period. At high frequencies the electrodes and the interelectrode space behave as a capacitor. The current created with alternative voltages passes across capacitors. Therefore, plasmas can be created inside spaces fully bordered by dielectric materials, leading to peculiar discharge configurations, as exemplified in Fig. 7.2, where dielectric barrier discharges (DBDs) with inductive coupling and annular electrodes are presented. Inside the discharge volume the electrical charges will move and oscillate with the periodic changing of the electrical field direction. At high frequencies (megahertz, for example) the ions, due their inertia, cannot follow the change in the electric field orientation. It is said that they are frozen in plasma.

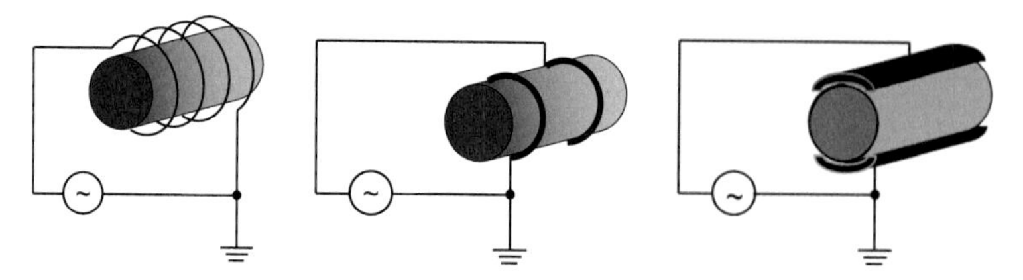

Figure 7.2 Inductive (left) and capacitive coupling: (center) with annular electrodes and (right) with parallel electrodes).

With respect to gas pressure, electrical discharges in gases can be created in a large range of pressure values. Until recently most plasma processing applications and the most important plasma sources were based on low-pressure discharges. Their advantages come from the following facts: (i) the purity of the plasma is very well controlled—usually a high vacuum is made first (10^{-5}–10^{-6} mbar), and after that pure gases (purity better that 99.9999% is usual for inert gases) are introduced at pressure values in the range of 10^{-2}–10 mbar; (ii) the electrical breakdown of the gas is easy to perform, because at a low pressure, the product *pd* is close to the minimum of Paschen curve at distances in the range of tens of centimeters, which are normal sizes of processing chambers; (iii) low-pressure discharges lead to volumetric plasmas; they extend over large volumes around the interelectrode space; in case the vacuum chamber is metallic, it can be used as a ground electrode and the entire chamber is filled with plasma; (iv) most low-pressure plasmas are cold plasmas. The gas temperature is closed to the room temperature, and processing of materials sensitive to temperature is possible without damage (e.g., polymers and biological materials); (v) low-pressure plasmas are quite stable; (vi) bombardment with ions of surfaces is possible due to acceleration of ions in the sheaths around the substrates; this is advantageous for cleaning or erosion of surfaces and for sputtering applications.

However, there are important disadvantages that limit the applicability of low-pressure plasma processing. For example, (i) vacuum equipment (mechanical, turbomolecular, ionic pumps, vacuum gauges, chambers, etc.) is expensive and its operation requires well-trained technical staff, (ii) the size of objects to be processed should be as per the dimensions of the vacuum chambers, (iii) the processing of porous materials (foams, textiles) is slow due to the long pumping down times, and (iv) some materials, for example, liquids or living biological entities, cannot be processed because of their incompatibility with vacuum.

7.2 Atmospheric Pressure Plasma Sources: Principles, Design, and Models

The last decades (starting with 1990) were marked by intense research aimed at developing plasma sources and systems working in ambient atmosphere, and therefore able to surpass the above disadvantages, while still trying to preserve the advantages of low-pressure processing. An analysis of the behavior of low-pressure gas discharge upon conditions (current, current density, pressure) aiming to clarify its extension as a non-equilibrium glow at high (atmospheric) pressure was performed in Ref. [2]. The major problem encountered in the course of development of atmospheric pressure, non-equilibrium plasma sources is related to the physical constraints limiting the production of plasmas characterized simultaneously by high pressure, large volume, and cold conditions. For a long time it was considered that except for corona discharge, nonthermal atmospheric pressure plasma sources are not possible because increasing the pressure in discharge causes the plasma to become hot due to its tendency to transition to arc ($T_g \approx T_e \approx T_i \approx 10^4$ K). Indeed, the current density in constant current normal glow discharges increases with the square of pressure, approaching rapidly the arc transition threshold. Near the threshold, electronic or thermal fluctuations lead to instability. Any local increase in temperature favors the local production of electrons, so more current will pass through that region, which leads to further heating, and this cycle goes on until the discharge regime switches to arc. Besides, this leads to plasma constriction (decrease of plasma sectional area, filamentation), which limits the discharge volume.

Various solutions were looked for in order to solve the problem of glow-to-arc transition, all of them being based on the same principle: the minimization of system heating. The successful approaches were based on the use of gases with high efficiency of heat transfer, like helium; the use of high-mass gas flow rates, which ensure fast transport of heat out of discharge; the use of pulsed and high-frequency voltages to control the power injected into the plasma at levels low enough to prevent a transition to arc; the use of dielectric barriers; and active cooling of the electrodes. In addition, the design of atmospheric pressure plasma sources had to solve the problem of breakdown. At atmospheric pressure the *pd* minimum falls in the submillimeter range. Therefore, most of the atmospheric pressure plasma sources use discharges in narrow spaces; otherwise very high voltages are necessary to switch on the discharge.

One of the first papers effectively demonstrating the existence of a cold homogeneous glow discharge at 1 atm is by Okazaki et al. [1, 2].

From a fundamental point of view the pioneering works on the mechanisms in homogeneous atmospheric pressure plasma were performed by Massines et al. [3]. They confirmed the existence of homogeneous cold atmospheric pressure dielectric discharge, showed the way to control it, and described its main behavior. Details on this topic can be found in the review in Ref. [4].

7.2.1 DBD and DBE Plasma Sources

A simple classification of plasma sources is made upon source geometry, axisymmetric or planar. More complex geometries are also possible. Another classification accounts for the discharge type. Thus, the sources are using DBD (dielectric barrier discharges) and DBE (discharge with bare electrodes) configurations. In the case of DBE sources, the plasma comes in direct contact with the electrodes, while in the case of DBD sources, it is separated from the electrodes by dielectric barriers. These classifications are exemplified in Figs. 7.3 and 7.4.

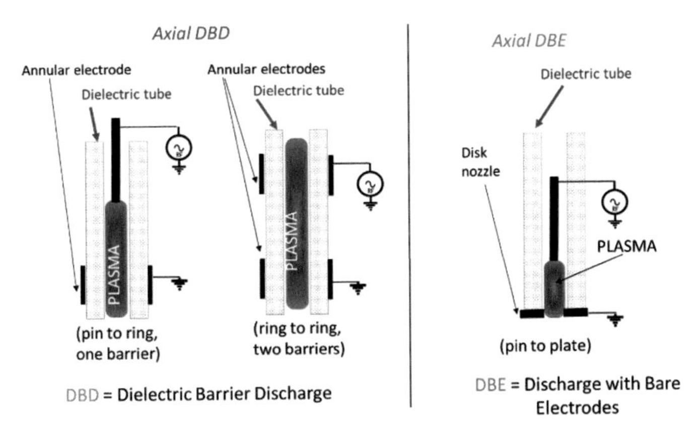

Figure 7.3 Schematic view of axisymmetric DBD (left) and DBE plasma sources (right).

Upon the discharge regime, atmospheric plasma can be generated in the homogeneous mode (glow) or in the filamentary or constricted mode. Coexistence of both modes is frequent. In well-controlled conditions, for example, at low power, homogeneous discharges can be obtained also at atmospheric pressure [5]. However, in some applications, the homogeneity is not a need and plasma sources can be operated in the easier-to-maintain filamentary mode.

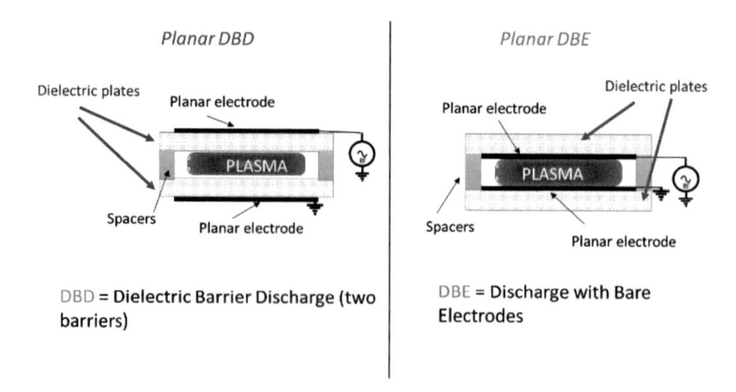

Figure 7.4 Schematic view of planar DBD and DBE plasma sources.

7.2.2 Expanding Plasmas, Plasma Jets

Finally, in processing applications, one difficulty to be surpassed is to ensure the contact of plasma with the material to be treated. Because of the narrow interelectrode space, the objects to be processed cannot be introduced directly in the atmospheric pressure discharges, except when they can be made parts of discharge, acting themselves as electrodes or barriers. The general solution to this problem is to extract the plasma and to use it outside the interelectrode space, as an expanding plasma jet [6]. Examples of configurations of plasma jets are presented in Figs. 7.5 and 7.6, and images of plasma jets are presented in Figs. 7.7 and 7.8.

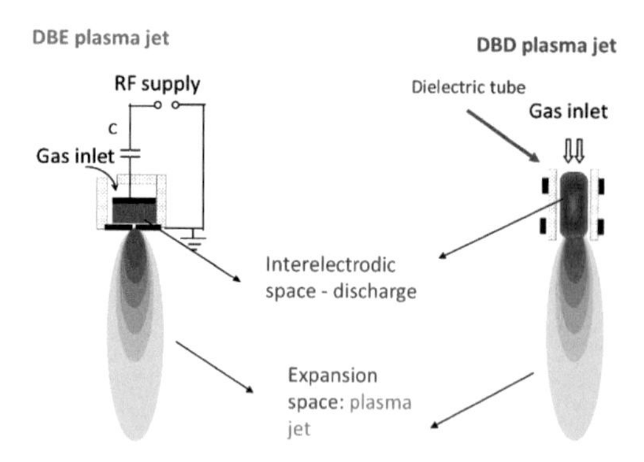

Figure 7.5 Schematic view of axisymmetric DBE and DBD plasma sources.

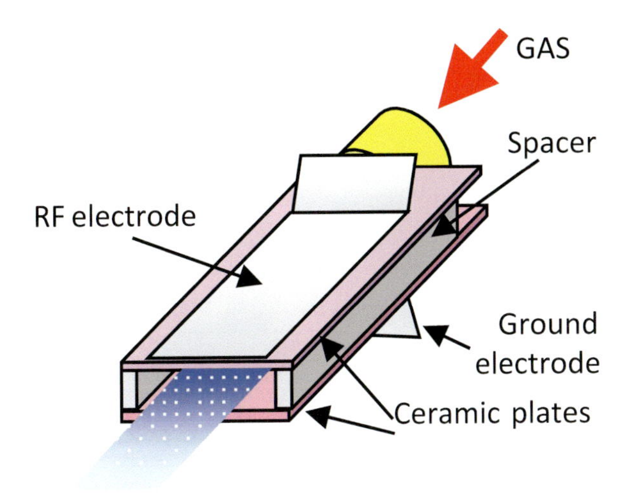

Figure 7.6 Schematic view of a planar DBD plasma source.

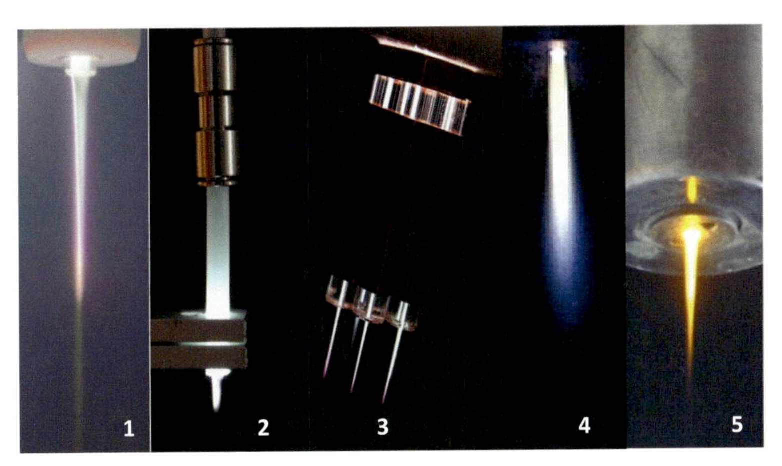

Figure 7.7 Images of axisymmetric plasma sources: (1) monoelectrode filamentary DBD Ar plasma jet [6], (2) diffuse DBD Ar plasma jet (annular electrodes), (3) multifilamentary DBD Ar plasma jets, (4) DBE nitrogen plasma jet [7], and (5) a DBE Ar plasma jet after immersion in a NaCl solution.

More details on plasma jet geometries can be found in the review paper in Ref. [8].

Figure 7.8 Images of planar plasma sources: (1) single-barrier DBD plasma jet, (2) DBE plasma jet with a transparent electrode (ITO), (3) double-barrier DBD plasma jet, and (4) double-barrier large-width DBD plasma jet during textile treatment at atmospheric pressure.

7.3 Applications of Atmospheric Pressure Plasma Sources

After 1992, devices started to appear like a microbeam plasma generator [9], a development named in later papers as a cold plasma torch. The processing potential of this device was illustrated in the following years in a series of publications devoted to applications like deposition of SiO_2 films [10, 11], deposition of TiO_2 films [12], ashing of photoresist [13], and etching [14].

On the way to develop practical devices, a milestone was the publication of the review paper (Hicks at University of California and Selwyn at Los Alamos) [15] where the main characteristics of a cold plasma jet device were compared with other atmospheric pressure sources, like corona discharge and transferred arc. In time a number of reviews were published on this topic concerning the devices, plasma diagnostics, and their applications [5, 8, 16].

In practical situations a source is built on a discharge configuration designed in connection with the application in view.

For local treatments, small size, axially symmetric plasma sources are appropriate. Contour processing can be performed by mounting the source on x-y stages and adding scanning procedures. For foil treatments large-area processing with planar DBD discharges or planar plasma jets might be a choice; efficient local treatment in linear geometry combined with a high transport velocity of the foil is another choice. In the next, the application potential of atmospheric pressure plasma sources is exemplified by surface modification, cleaning of carbon layers from irregular surfaces, thin film deposition at atmospheric pressure, and decomposition of chemicals from liquids. These applications make use of RF plasma sources, sustained at 13.56 MHz. The RF discharges present advantages: designs based on dielectric barriers are accessible, the dependence of the breakdown voltages upon frequency presents a minimum around 13.56 MHz, and because of their extensive use in the semiconductor industry, the RF generators are among the most advanced power supplies with automatic power control.

7.3.1 Surface Modification

Two basic properties of surfaces are the roughness and the chemical composition. These properties play a key role in many phenomena and applications, like wettability, dyeing, adhesion, and interaction with biological molecules and living entities. There are specific techniques that allow surface characterization from these points of view: for roughness characterization atomic force microscopy (AFM) and profilometry are mostly used, while the type of chemical elements present at the surface is revealed by X-ray photoelectron spectroscopy (XPS) and energy-dispersive X-ray spectroscopy (EDS). Even more, details on the bonding types between the chemical elements can often be extracted from XPS data.

Wettability phenomena refer to the interaction of surfaces with liquids. The easiest-to-understand example is water wettability. In practice water wettability is measured by the sessile drop method. In this method a small drop of water (typically a few microliters, meaning a drop with a diameter of around 1 mm) is gently deposited on the solid surface and the angle between the water surface and solid surface at contact is measured. With respect to the value of the contact angle, the solid surfaces are classified as hydrophilic if the water contact angle (WCA) is less than 90° and hydrophobic if it is larger than this value. Inside these ranges surfaces with extreme wettability are found: superhydrophilic (WCA less than 10°) and superhydrophobic (WCA larger than 150°).

The physics behind the wettability phenomena relates to the interaction of water molecules with those of the solid at the solid/liquid interface. Basically, because the water molecules are polar, the solids with polar groups (or charges) at the surface present a hydrophilic character. If the interaction is low, then the surfaces are hydrophobic. The roughness of the surface also influences the wettability; it amplifies the intrinsic wettable character of a material: the roughness increases the geometrical area of contact, and the solids, which present a hydrophilic character as smooth surfaces, exhibit a more hydrophilic character as rough surfaces, while the smooth surface hydrophobic materials show a more hydrophobic character when their surfaces become rough.

Atmospheric pressure plasmas are important processing tools for handling the surfaces' wettability because of their ability to modify both the surface chemistry and the surface roughness. This is particularly significant in polymer processing. The reactive species generated in plasmas will act on the polymeric surface, breaking bonds and creating defects. Plasma-generated polar radicals link to these chemically active sites, rendering to the surface a polar component, thus normally making it more hydrophilic than the untreated surface. The opposite can also be done by plasma fluorination: it is well known that fluorine containing polymers are highly hydrophobic, a common example being Teflon (polytetrafluorethene). On being treated with fluorine containing plasmas, polymer surfaces become more hydrophobic than the initial ones. The fluorine atoms and radicals created in the plasma bind at surfaces and hinder the existing charges, making the surface very inert with respect to the interaction with water molecules.

The second aspect is related to surface roughness. Interaction of plasma species with the polymer surface will lead to the removal of some of the atoms, or even small fragments of material, and therefore to erosion. The eroded material is released in the ambient as volatile components or as small solid fragments. Following this process the development of roughness at the surface is noted, which together with the plasma-induced chemistry modification contributes to emphasizing the new character of wettability.

A simple experiment illustrating the modification of wettability by plasma is presented in Fig. 7.9. In this experiment the polymer substrate is covered by a mask having regular circular openings of 100 μm and it is exposed to an atmospheric plasma jet through the mask. During treatment, only the substrate zones corresponding to the mask openings come in contact with plasma species and are therefore modified. After removing the mask, one can promote ambient water vapor condensation on the surface by cooling the substrate, using a Peltier element or even placing the substrate on a cold surface. The interaction of the water vapors with the surface is stronger on the hydrophilic zones, where the water molecules get linked on the modified substrate. Water accumulates on these zones, in contrast with the hydrophobic, untouched by plasma, zones. Therefore, regular arranged waterdrops, mimicking the mask openings, are formed on the substrate, as shown in Fig. 7.9c.

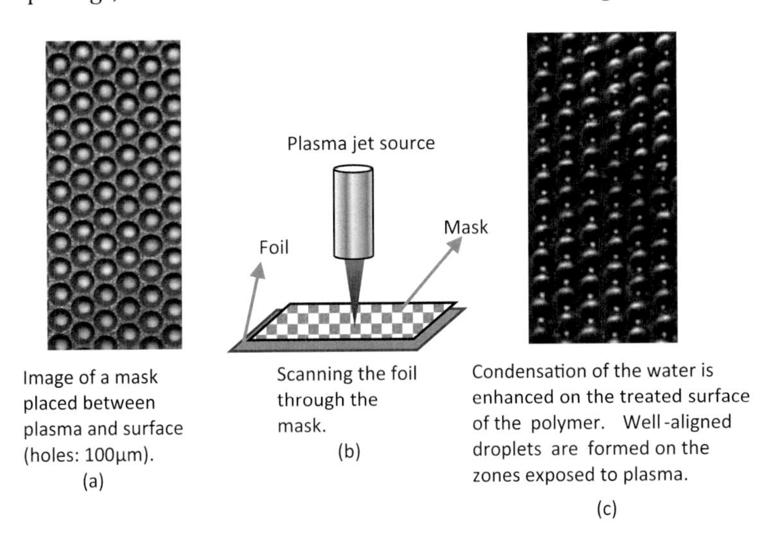

Figure 7.9 Illustration of plasma-induced hydrophilicity of polymer surfaces.

Other questions related to polymer surface changes caused by plasma are how fast they proceed and for how long they persist. The change in the chemical character of the surface in contact with plasma is very fast. A change in the contact angle at a rate of 1° per millisecond can be measured. The roughness modification is easy to notice by AFM measurements after a few minutes of treatment, the values depending on the plasma and polymer type.

More details on polymer surface modifications are presented in Ref. [17–19].

7.3.2 Coating of Surfaces by Deposition with Atmospheric Pressure Plasma Jets

A general classification of gas phase deposition techniques divides them into physical vapor deposition (PVD) and chemical vapor deposition (CVD) categories. These are atomistic techniques because the deposited layers build up from atoms and molecules. In CVD techniques a gaseous precursor is fragmented in the vicinity of the substrate and the resultant fragments react at the surface, forming a solid component (the thin film) and a volatile one, which is removed by the gas flow. The fragmentation process and the surface reaction require energy, which among others can be provided by heating (thermal CVD), plasma (plasma-assisted CVD), and light (photo-CVD). Low-pressure plasma-assisted CVD is widely used (deposition of thin films in semiconductor industry, architectural coatings, hard coatings on tools, etc.), while atmospheric pressure plasma CVD is in the course of development.

Here, we illustrate the principle of deposition at atmospheric pressure with contour deposition of carbon material [20]. In this experiment a small size plasma jet source is used, working in argon, and acetylene is injected laterally through an injection chamber (schematic view in Fig. 7.10). Following the mixing and interactions with plasma electrons and excited argon species, acetylene molecules are decomposed in radicals that reach the substrate and lead to a local deposit of carbon material. A predefined contour (in the present case a spiral and a small square) is then obtained by moving the plasma source with a scanning stage controlled by a computer.

For large-area processing, planar plasma sources may be used. A solution of symmetric injection is possible by using a design where two plasma layers are created and injection is performed between them [21].

More details on deposition with atmospheric pressure plasma sources can be found in Ref. [22].

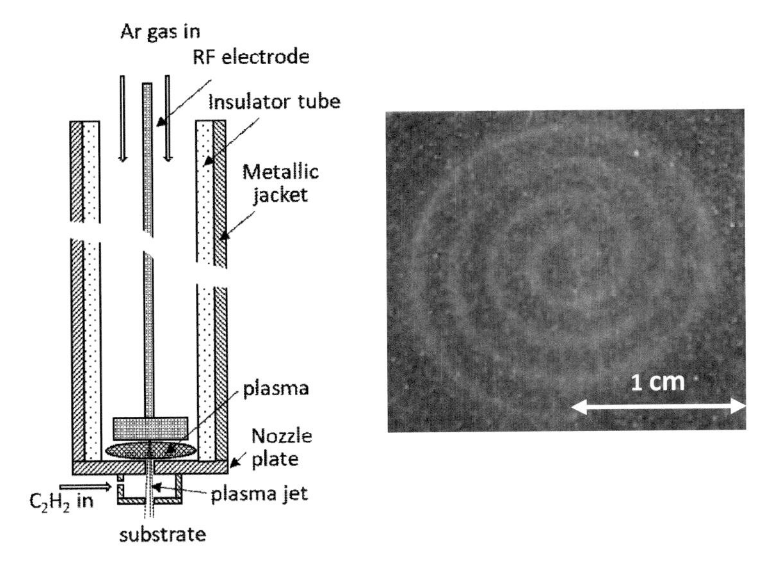

Figure 7.10 (a) Schematic view of an atmospheric pressure plasma jet source (DBE) with a lateral injection of precursor (acetylene) used for carbon material deposition. (b) A spiral contour performed by the motion of the plasma source above the substrate.

7.3.3 Surface Cleaning with Atmospheric Pressure Plasma Jets

Removal of contaminants and of material deposits from surfaces is necessary in many applications. For example, mirrors of very high reflectivity are used in large research infrastructures using high-power lasers and in synchrotrons. Any decrease in reflectivity may compromise the mirrors by inducing the destruction of the surface by local absorption of the light beams. Keeping the surface clean is a must in such situations. Surface contamination is, however, inherent, even in high vacuum, due to the presence of volatile organic compounds in the environment: they condense on surfaces, forming a very thin hydrocarbon film. This film should be removed in order to preserve the mirror properties. In case of tokamaks, erosion of walls during plasma phase releases material as vapors, particulates, and debris fragments, covering over time the mirrors and various chamber wall regions with redeposited layers. Besides compromising the mirror functionality, the redeposited layers should be removed because they are trapping the fuel (tritium). Other cleaning applications include the removal of the plastic residues from injection molds and removal of biofilms from medical devices.

A solution to these problems can be plasma cleaning. The basic process is based on the interaction of reactive species present in the plasma with the material to be removed. Formation and release of volatile radicals containing atoms of the contaminating layer is the key: these volatile species are transported away from the surface by the gas flow. A typical example is the removal of organic layers from surfaces by oxygen, nitrogen, and hydrogen containing plasmas. These plasmas are rich in radicals that easily react with carbon, forming volatile molecules like CO, CN, and CO_2. In the case of low-pressure plasma, these chemical processes are enhanced by ion bombardment, which stimulates the volatile species desorption. The substrate temperature also influences the reactions. Even the ion bombardment is less active when surfaces are exposed to atmospheric pressure plasmas, they still present important advantages: they can be used in an ambient atmosphere, can produce local cleaning due to their small size, and can be adapted to large-area processing by scanning.

Here, we illustrate the plasma cleaning process with an example of the removal of carbon residues from flat [23] and complex-shape surfaces [24] (Fig. 7.11) with an atmospheric pressure plasma jet. The example has relevance for tokamak wall cleaning. To withstand the mechanical stresses caused by the high fluxes of heat and energetic particle fluxes, the tiles in tokamak are castellated, being assembled from pieces separated by gaps of 0.5–2 mm. In the case of tokamaks with carbon divertors, carbon layers are redeposited on the tiles and in the gaps. We simulated such complex surfaces by assembling small cubes coated on all faces with 2 μm thick layers of hydrogenated carbon. The coatings were produced by plasma-assisted CVD in a low-pressure argon RF discharge using acetylene as the precursor. The cleaning effect is clearly evidenced after scanning with a plasma jet along the gaps delimiting the central cube (Fig. 7.7). Removal rates in the range of a 2–10 mg/min were obtained with a 3000 sccm nitrogen plasma jet operated at 400 W with the addition of up to 1000 sccm of oxygen [24].

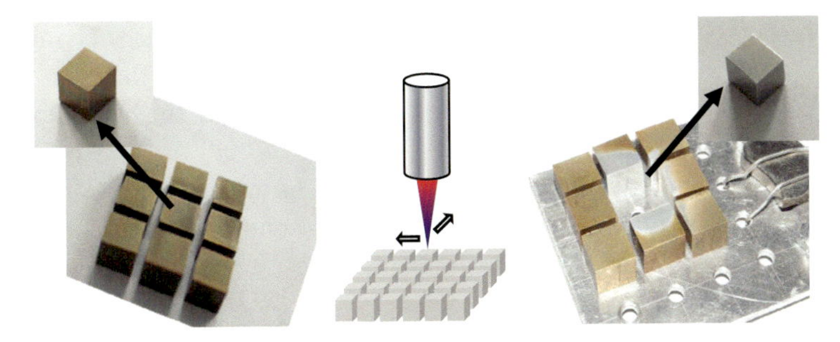

Figure 7.11 Illustration of the plasma cleaning effect after scanning with the plasma jet along the gaps delimiting the cubes. Complete removal of the material from inside gaps and front face is achieved. Reprinted from Ref. [24], Copyright (2016), with permission from Elsevier.

7.3.4 Atmospheric Pressure Plasma Processing of Liquid Solutions and Dispersions

Plasma processing of liquid solutions and dispersions requires the contact of the plasma species with the liquid. The literature is rich with examples on discharge configurations suitable for operation with liquids [25]; here we restrict the discussion to cold plasma jets. The experimental approaches are based on plasma operation either in contact with or submerged inside liquids.

Figure 7.12 Inside-liquid plasma treatment with a filamentary plasma jet.

A representative example is shown in Fig. 7.12, where an image of an inside-liquid plasma treatment with a filamentary jet is presented. In this case liquid penetration in the discharge channel is prevented by allowing the feeding gas to overflow through the discharge, thus counterbalancing, at the nozzle level, the hydrostatic pressure of the liquid.

The range of applications is quite large. The main subjects concern degradation of chemical compounds in solutions, nanomaterial synthesis and modification in liquid phase, and bacterial inactivation.

7.3.4.1 Degradation of chemical contaminants in solutions

Specifically when mixed with a liquid, the plasma species changes the liquid's chemical composition; for example, in aqueous liquids, hydrogen peroxide molecules (O_2H_2) and OH radicals are created. They have a high potential to oxidize, leading to the degradation of chemical substances in solutions or bacterial inactivation. In-liquid plasma can drive various chemical processes, depending on conditions. The decomposition of chemicals is of wide interest for water cleaning and pollutant decomposition. Such procedures have additional benefits related to the use of low temperatures during treatment, relatively simple installations (without vacuum components), and low operating cost. Still, the ways for maximizing the number of reactive species, range of operational parameters, long-time stability, and suitability of plasma sources for such applications should be studied.

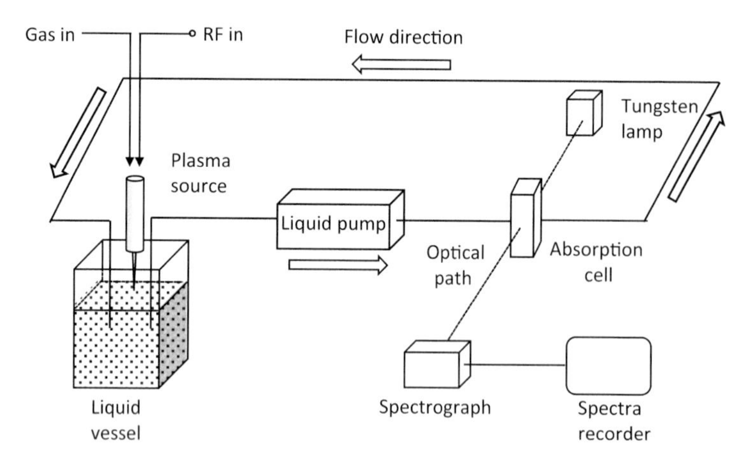

Figure 7.13 Schematic view of an experimental setup used in the plasma degradation of dyes in a solution.

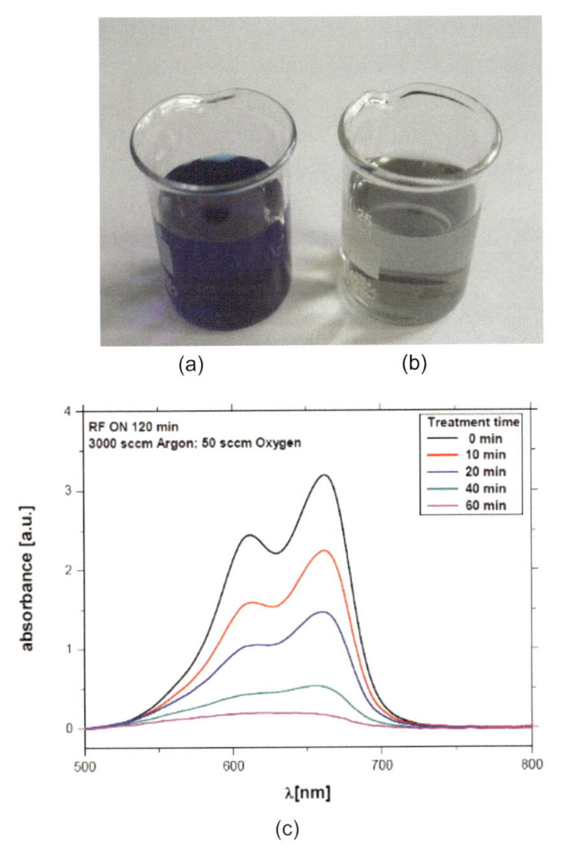

Figure 7.14 Recipients containing a methylene blue solution prior to (a) and after (b) plasma treatment and a plot (c) showing the absorbance evolution with treatment time.

The typical examples illustrating the potential of degradation of chemical compounds with liquids use organic dyes as substances. For dyes the degradation process is easily monitored by recording the decolorization with the use of UV-visible absorption spectrophotometry. This technique directly relates the experimentally measured solution absorbance with the concentration of the absorbing substance through the Beer–Lambert law. A schematic view of an experimental setup used in a degradation experiment is presented in Fig. 7.13. The main parts of the setup are the plasma source with the treatment cell, the liquid circulation system, and the UV-visible monitoring system. To perform the experiments, a cold (nonthermal) atmospheric pressure small-size plasma jet was generated in argon by a low-power RF discharge [26]. The plasma jet was positioned above the liquid in proximity to the surface or submerged into the solution. The effect of the gaseous atmosphere on the degradation of methylene blue could be investigated by allowing methylene blue to flow in the reaction chamber filled with different gases (such as N_2, O_2, and air). The modification of the chemical composition of the solution upon treatment was easily observed visually (Fig. 7.14a,b). Also, the absorbance spectra showed a significant decolorization of the prepared solution as the treatment time increased (Fig. 7.14c). Among the parameters investigated in this experiment [26], an important one is the position of the plasma jet relative to the liquid surface. The comparison of decomposition characteristics indicates that the dye conversion due to the plasma jet is much faster and more energy efficient when the plasma is submerged into the solution. Such types of experiments belong to the very intense research field devoted to water decontamination and purification, more details of which can be found in Ref. [27].

7.3.4.2 Plasma in-liquid processing of nanomaterials

Examples of nanomaterials include nanopowders consisting of nanoparticles, graphene carbon sheets, and natural nanomaterials like clay nanosheets and cellulose nanofibrils. For such materials applications may involve an improvement in the rheological properties (flowability, prevention of agglomeration, etc.); surface modification to promote interaction with molecules, cells, and drugs; and deposition of thin films on the surface to make core-shell particles. Gas phase processing is not adequate for such applications because of the danger of contamination of the atmosphere and the difficulty in ensuring direct contact of the plasma species with every particle. Instead, by using plasma in liquid, dispersions of nanopowders are first made and then these dispersions are plasma-treated, with effects on dispersion properties (sedimentation rate, viscosity, etc.) and also on material properties (morphology, structure, size, surface chemistry, etc.). A few examples concerning plasma modification of graphene nanosheets and cellulose nanofibrils can be found in Refs. [28–30]. More details can be found in the review papers in Refs. [31, 32].

7.3.5 Other Applications

Some of the most investigated topics are the use of cold plasmas in biology and medicine. Atmospheric pressure plasmas present an antimicrobial effect when applied to surfaces, liquids, or gels. An early paper discussing this effect is the review paper of Laroussi [33]. In the paper the author concludes that "Insights into the roles of UV radiation, active species, and charged particles has led to the conclusion that chemically reactive species, such as free radicals, play the most important role in the inactivation process." More details on the mechanism and results can be found in Ref. [34]. On the basis of their antimicrobial effect, atmospheric pressure plasmas have found applications in food preservation. Plasma sterilization aspects related to food technology can be found in Refs. [35, 36].

A very active field is plasma medicine. Studies are devoted to healing of chronical wounds, blood coagulation, surgery, and cancer treatment. Details can be found, for example, in the review paper in Ref. [37] and the book in Ref. [38].

7.4 Conclusions and Outlook

At present, most of cold atmospheric pressure plasma fundamentals are established. A large number of plasma sources with various configurations were demonstrated. However, the wide applicability is still limited by a few drawbacks, like:

- The plasma size: Cold, large-area atmospheric pressure plasma sources are in high demand and critical for many plasma technologies. Atmospheric plasma is typically small; plasma jets evolve through apertures (nozzles) a few (1–2) millimeters in diameter. Scaling up of the sources in order to increase the footprint of the plasma at the substrate was in view of different groups along the time, and it is a very hot topic. Planar plasma sources, like that presented in Fig. 7.8, can be extended to tens of centimeters in length. In the works of Konesky [39, 40] a planar or flexible jet array is demonstrated that uses an assembly of 22 plasma jets, the current in each of them being limited by a separate ballast resistance. On the same principle, a plasma jet array of 45 jets, assembled to cover an area of 50 mm × 150 mm, is reported by the same author in Ref. [41]. The plasma source is operated with helium gas and has the electrodes in contact with the plasma. An array of seven plasma jets grouped in a honeycomb structure a few centimeters in diameter was reported in Refs. [42, 43]. An innovative idea is the design of a "wearable atmospheric pressure plasma system" that proposes an open-air device consisting of interdigitated wires assembled in a fabriclike electrode structure [44].

- Erosion of electrodes: When the plasma comes in contact with the electrodes erosion of electrodes is noticed over time and replacement might be necessary. Physical erosion is driven by bombardment with charged species; chemical erosion is sustained by reactive plasma radicals. This can be avoided in the configurations with the electrodes placed outside the plasma generation, as DBDs. Also, electrode damage is drastically diminished by operating the sources in the glow discharge mode.
- Nature of gas and gas consumption: Most of the plasmas working at atmospheric pressure use high gas flow rates in order to ensure cooling and discharge stabilization; after excitation in the interelectrode discharge the gas is ejected out of a nozzle as a plasma jet. Most publications report the use of helium as the working gas. Helium is an expensive gas, and a specific gas recovery system might be necessary. A trend to work in industrial argon or even in air is noticed.

In spite of these limiting factors cold atmospheric plasma technology has reached a level of maturity in surface modification and sterilization. Also, a number of plasma medical devices are on the market. A number of companies are producing cold atmospheric plasma devices and systems both for research and industrial activities.

References

1. S. Kanazawa, M. Kogoma, T. Moriwaki, S. Okazaki, Stable glow plasma at atmospheric pressure, *J. Phys. D: Appl. Phys.*, **21** (1988) 838–840.

2. S. Okazaki, M. Kogoma, M. Uehara, Y. Kimura, S. Okazaki, Appearance of stable glow discharge in air, argon, oxygen and nitrogen at atmospheric pressure using a 50 Hz source, *J. Phys. D: Appl. Phys.*, **26** (1993) 889.
3. F. Massines, A. Rabehi, P. Decomps, R. B. Gadri, P. Segur, C. Mayoux, Experimental and theoretical study of a glow discharge at atmospheric pressure controlled by dielectric barrier, *J. Appl. Phys.*, **83**(6) (1998) 290–295.
4. P. J. Bruggeman, F. Iza, R. Brandenburg, Foundations of atmospheric pressure non-equilibrium plasmas, *Plasma Sources Sci. Technol.*, **26**(12) (2017) 123002.
5. P. J. Bruggeman, F. Iza, R. Brandenburg, Foundations of atmospheric pressure non-equilibrium plasmas, *Plasma Sources Sci. Technol.*, **26**(12) (2017) 123002.
6. G. Dinescu, E. R. Ionita, Radio frequency expanding plasmas at low, intermediate, and atmospheric pressure and their applications, *Pure Appl. Chem.*, **80**(9) (2008) 1919–1930.
7. G. Dinescu, S. Vizireanu, C. Petcu, B. Mitu, M. Bazavan, I. Iova, Spectral characteristics of a radiofrequency nitrogen plasma jet continuously passing from low to atmospheric pressure, *J. Optoelectron. Adv. Mater.*, **7**(5) (2005) 2477–2480.
8. F. Fanelli, F. Fracassi, Atmospheric pressure non-equilibrium plasma jet technology: general features, specificities, and applications in surface processing of materials, *Surf. Coat. Technol.*, **322** (2017) 174–201.
9. H. Koinuma, H. Ohkubo, T. Hashimoto, K. Inomata, T. Shiraishi, A. Miyanaga, S. Hayashi, Development and application of a microbeam plasma generator, *Appl. Phys. Lett.*, **60**(7) (1992) 816–817.
10. K. Inomata, H. Ha, K. A. Chaudhary, H. Koinuma, Open air deposition of SiO_2 film from a cold plasma torch of tetramethoxysilane-H_2–Ar system, *Appl. Phys. Lett.*, **64**(1) (1994) 46–48.
11. S. E. Babayan, J. Y. Jeong, V. J. Tu, J. Park, G. S. Selwyn, R. F. Hicks, Deposition of silicon dioxide films with an atmospheric pressure plasma jet, *Plasma Sources Sci. Technol.*, **7**(3) (1998) 286–288.
12. H.-K. Ha, M. Yoshimoto, H. Koinuma, Opcn air plasma chemical vapor deposition of highly dielectric amorphous TiO_2 films, *Appl. Phys. Lett.*, **68** (1996) 2965.
13. K. Inomata, H. Koinuma, Open air photoresist ashing by a cold plasma torch: catalytic effect of cathode material, *Appl. Phys. Lett.*, **66**(17) (1995) 2188–2190.

14. Y. Jeong, S. E. Babayan, V. J. Tu, J. Park, R. F. Hicks, G. S. Selwyn, Etching materials with an atmospheric-pressure plasma jet, *Plasma Sources Sci. Technol.*, **7**(3) (1998) 282–285.
15. A. Schutze, J. Y. Jeong, S. E. Babayan, J. Park, G. S. Selwyn, R. F. Hicks, The atmospheric-pressure plasma jet: a review and comparison to other plasma sources, *IEEE Trans. Plasma Sci.*, **26** (1998) 1685–1694.
16. C. Tendero, C. Tixier, P. Tristant, J. Desmaison, P. Leprince, Atmospheric pressure plasmas: a review, *Spectrochim. Acta, Part B*, **61**(1) (2006) 2–30.
17. M. D. Ionita, M. Teodorescu, T. Acsente, M. Bazavan, E. R. Ionita, G. Dinescu, Remote surface modification of polymeric foils by expanding atmospheric pressure radiofrequency discharges, *Rom. J. Phys.*, **56** (2011) S132–S138.
18. M. D. Ionita, M. Teodorescu, C. Stancu, C. E. Stancu, E. R. Ionita, A. Moldovan, T. Acsente, M. Bazavan, G. Dinescu, Surface modification at atmospheric pressure in expanding RF plasmas generated by planar dielectric barrier discharges, *J. Optoelectron. Adv. Mater.*, **10**(3) (2010) 777–782.
19. E. R. Ionita, M. D. Ionita, C. E. Stancu, M. Teodorescu, G. Dinescu, Small size plasma tools for material processing at atmospheric pressure, *Appl. Surf. Sci.*, **255**(10) (2009) 5448–5452.
20. G. S. Vlad, E. R. Ionita, I. Ciobanu, C. M. Petcu, G. Dinescu, Processing of selective contours on flat surfaces by computer assisted plasma beam tracking, in M. Mutlu, G. Dinescu, R. Forch, J. M. Martin-Martinez, J. Vyskocil (eds.), *Plasma Polymers and Related Materials*, Hacettepe University Press (2005).
21. G. Dinescu, M. Teodorescu, C. Stancu, E. R. Ionita, Cold plasma source with two planar jets and gas injection for thin film deposition, etching, cleaning and functionalization of surfaces at atmospheric pressure, patent A/00864/2015 (Romania) (2015).
22. H. Kakiuchi, H. Ohmi, K. Yasutake, Atmospheric-pressure low-temperature plasma processes for thin film deposition, *J. Vac. Sci. Technol., A*, **32** (2014) 030801.
23. E. R. Ionita, I. Luciu, G. Dinescu, C. Grisolia, Flexible small size plasma torch for Tokamak wall cleaning, *Fusion Eng. Des.*, **82** (2007) 2311–2317.
24. C. Stancu, D. Alegre, E. R. Ionita, B. Mitu, C. Grisolia, F. L. Tabares, G. Dinescu, Cleaning of carbon materials from flat surfaces and castellation gaps by an atmospheric pressure plasma jet, *Fusion Eng. Des.*, **103** (2016) 38–44.

25. P. Vanraes, A. Bogaerts, Plasma physics of liquids: a focused review, *Appl. Phys. Rev.*, **5** (2018) 031103.

26. E. C. Stancu, D. Piroi, M. Magureanu, G. Dinescu, Decomposition of methylene blue by a cold atmospheric pressure plasma jet source, *Proceedings of the 20th International Symposium on Plasma Chemistry*, July 24–29, 2011, Philadelphia, USA (2011).

27. J. E. Foster, Plasma-based water purification: challenges and prospects for the future, *Phys. Plasmas*, **24** (2017) 055501.

28. M. D. Ionita, S. Vizireanu, S. D. Stoica, M. Ionita, A. M., Pandele, A. Cucu, I. Stamatin, L. C. Nistor, G. Dinescu, Functionalization of carbon nanowalls by plasma jet in liquid treatment, *Eur. Phys. J. D*, **70** (2016) 31.

29. S. Vizireanu, D. M. Panaitescu, C. A. Nicolae, I. Chiulan, M. D. Ionita, V. Satulu, L. G. Carpen, S. Petrescu, R. Birjega, G. Dinescu, Cellulose defibrilation and functionalization by plasma in liquid treatment, *Sci. Rep.*, **8** (2018) 1543.

30. D. M. Panaitescu, S. Vizireanu, C. A. Nicolae, A. N. Frone, A. Casarica, L. G. Carpen, G. Dinescu, Treatment of nanocellulose by submerged liquid plasma for surface functionalization, *Nanomaterials*, **8**(7) (2018) 467.

31. Q. Chen, J. Li, Y. Li, A review of plasma-liquid interactions for nanomaterial synthesis, *J. Phys. D: Appl. Phys.*, **48** (2015) 424005.

32. S. Horikoshi, N. Serpone, In liquid plasma: a novel tool in the fabrication of nanomaterials and the treatment of wastewaters, *RSC Adv.*, **7**(75) (2017) 47196–47218.

33. M. Laroussi, Non-thermal decontamination of biological media by atmospheric pressure plasmas: review, analysis and prospects, *IEEE Trans. Plasma Sci.*, **30**(4) (2002) 1409–1415.

34. J. W. Lackmann, J. E. Bandow, Inactivation of microbes and macromolecules by atmospheric-pressure plasma jets, *Appl. Microbiol. Biotechnol.*, **98**(14) (2014) 6205–6213.

35. B. A. Niemira, Cold plasma decontamination of foods, *Ann. Rev. Food Sci. Technol.*, **12**(3) (2012) 125–142.

36. J. Pinela, I. C. Ferreira, Nonthermal physical technologies to decontaminate and extend the shelf-life of fruits and vegetables: trends aiming at quality and safety, *Crit. Rev. Food Sci. Nutr.*, **57**(10) (2017) 2095–2111.

37. G. Y. Park, S. J. Park, M. Y. Choi, I. G. Koo, J. H. Byun, J. W. Hong, J. Y. Sim, G. J. Collins, J. K. Lee, Atmospheric-pressure plasma sources for biomedical applications, *Plasma Sources Sci. Technol.*, **21** (2012) 043001.

38. X. P. Lu, S. Reuter, M. Larroussi, D. W. Liu, Nonequilibrium atmospheric pressure plasma jest, in *Fundamental, Diagnostics, and Medical Applications*, CRC Press, Taylor and Francis Group (2019).
39. G. Konesky, Large area cold plasma applicator for decontamination, in *Chemical, Biological, Radiological, Nuclear and Explosives (CBRNE) Sensing IX*; 69540E, eds. A. W. Fountain III, P. J. Gardner, Book Series Proceedings of the Society of Photo-Optical Instrumentation SPIE, Vol. 6954, pp. E9540–E9540 (2008).
40. Konesky, G. Cold plasma decontamination using flexible jet arrays, in *Chemical, Biological, Radiological, Nuclear and Explosives (CBRNE) Sensing XI*; 76651P, eds. A. W. Fountain III, P. J. Gardner, Book Series Proceedings of SPIE-The International Society for Optical Engineering, Vol. 7665, Article number 76651 (2010).
41. G. Konesky, Dwell time considerations for large area cold plasma decontamination, *Proc. Chemical, Biological, Radiological, Nuclear, and Explosives (CBRNE) Sensing X*, Vol. 7304, pp. 73040N-1 to -10 (2009).
42. Z. Cao, Q. Nie, D. Bayliss, J. L. Walsh, C. S. Ren, D. Z. Wang, M. G. Kong, Spatially extended atmospheric plasma arrays, *Plasma Sources Sci. Technol.*, **19** (2010) 025003.
43. J. E. Kim, J. Ballato, S.-O. Kim, Intense and energetic atmospheric pressure plasma jet arrays, *Plasma Processes Polym.*, **9**(3) (2012) 253–260.
44. H. Jung, J. A. Seo, S. Choi, Wearable atmospheric pressure plasma fabrics produced by knitting flexible wire electrodes for the decontamination of chemical warfare agents, *Sci. Rep.*, **7** (2017) 40746.

Chapter 8

Plasma in Odontology

Sara Laurencin-Dalicieux,[a] Marie Georgelin-Gurgel,[b] Jean Larribe,[c] Antoine Dubuc,[b] and Sarah Cousty[d]

[a]*Dental Faculty, Paul Sabatier University, Toulouse, France*
[b]*CHU Toulouse, Toulouse, France*
[c]*INSERM U1043, Université de Toulouse, Toulouse, France*
[d]*Lapace F-31062, Université de Toulouse, Toulouse, France*
cousty.s@chu-toulouse.fr

8.1 Introduction

What is odontology? Odontology, also called "dentistry," represents the profession concerned with the prevention and treatment of child and adult oral diseases. It includes not only diseases of the teeth and supporting structures but also diseases of the mouth's soft tissues, including the oral mucosa and salivary glands. Odontology also encompasses the treatment and correction of jaw malformations, temporomandibular joint disorders, teeth misalignment, and malocclusion. As a result, many specialties and subspecialties exist, such as periodontics, prosthodontics, oral surgery, oral medicine and pathology, endodontics, and pedodontics.

Plasma Applications for Material Modification: From Microelectronics to Biological Materials
Edited by Francisco L. Tabarés

ISBN 978-981-4877-35-0 (Hardcover), 978-1-003-11920-3 (eBook)
www.jennystanford.com

"Plasma in odontology" is one part of the wide field of plasma medicine.

The specificity of the oral cavity lies in the presence of a complex bacterial flora and saliva and surface biofilms. By the early 2000s, Stoffels had introduced the first investigation periodontal treatment with direct nonthermal plasma [1]. In fact, dental applications of gaseous plasma can be divided into two main approaches:

- Surface treatment of materials or medical devices
- Direct application in the mouth

We decided to present the different plasma applications in odontology using this "artificial" classification, even though we are well aware that many overlaps exist, in particular, with decontamination and sterilization (Fig. 8.1) processes.

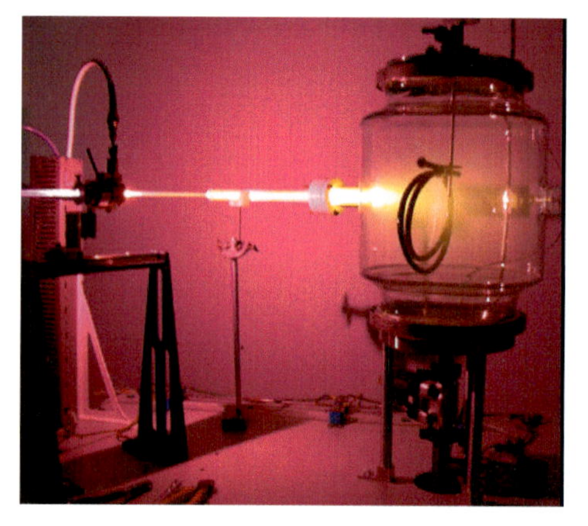

Figure 8.1 Example of a pure N_2 post discharge used for sterilization applications [2].

8.2 Surface Treatment of Materials or Medical Devices

8.2.1 Peri-implant Osseointegration Improved by Plasma Treatment

This might be the most famous application of plasma technology in dentistry. Dental implants are part of the standard procedure in the replacement of teeth in contemporary dentistry. The concept of osseointegration was first described by Branemark in the 1950s. It is the result of a cascade of events leading to bone healing and intimate interaction with the implant surface. Osseointegration influences the functional implant loading. Many approaches have been studied for enhancing the speed of osseointegration and, therefore, the implant stability. Wettability (hydrophilicity) of a titanium implant's surface plays a major role in cell adhesion. As a result, plasma glow discharges at low pressure have been used during the manufacturing process [3]. Hydrocarbon and functional hydroxyl groups can be added to the implant's surface, the advantage being that there is no residue after plasma treatment.

Zirconia implants are a good alternative to conventional titanium implants, thanks to their esthetic properties, biocompatibility, and reduced bacterial colonization. Also, increased hydrophilicity and osseointegration in in vitro and in vivo experiments has been reported after plasma treatment [4].

Recently, cold atmospheric plasmas (CAPs) have been used to modify the implant surface immediately prior to implant placement [5]. One of the advantages of this chair-side treatment is the absence of residue. Indeed, it is based on the modification of the surface's physicochemical properties (e.g., surface energy and functional hydroxyl groups) [6].

The authors found that plasma treatment reduced the contact angle and supported the spread of osteoblastic cells [7, 8].

Ujino et al. highlighted the importance of hydrophilicity and removing contaminants from the titanium implant surface, using plasma treatment to improve the initial adhesion of protein and cells [9].

All these variations in surface texture can be observed with plasma spraying. It is a well-known coating technique for implants in order to improve osteointegration (Fig. 8.2). Plasma thin coating techniques can be nitriding [11], titanium nitride oxide coatings, plasma-polymerized hexamethyldisiloxane [12], plasma-polymerized allylamine [13], and plasma-polymerized acrylic acid [14].

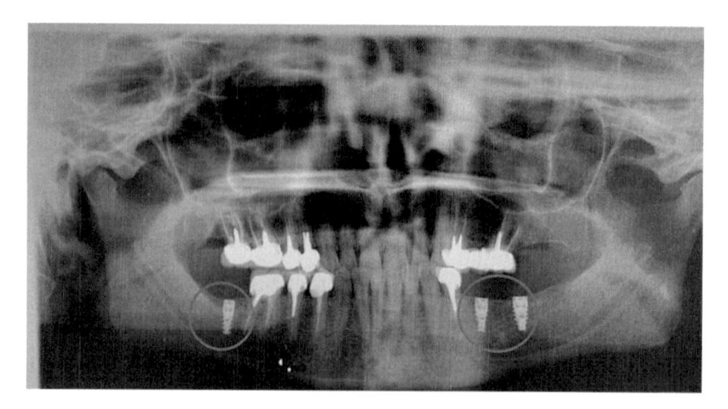

Figure 8.2 Panoramic radiograph—complete osseointegration of the implants [10].

8.2.2 Increasing the Adhesion

Adhesive dentistry is now the gold standard of restorative dentistry. Several systems are used to obtain optimal adhesion, which means optimal wettability; elevate surface energy; and increase roughness. Use of "etch-and-rinse" techniques, acid primers, and hydroxyethylmethacrylate primers is conventional. Recently, laser irradiation has been tested. Plasma treatment could be an additional procedure, or an alternative, in the bonding.

Today, the bonding of ceramic restorations is difficult to achieve and needs several steps that are complex, time intensive, and costly. By the way, this procedure may require toxic chemicals. The use of atmospheric pressure plasma could be of interest in this indication. Plasma technology produces carboxyl groups on the ceramic surface; as a result, hydrophilicity is improved [15].

Zirconia has a nonreactive and hydrophobic surface, with poor adhesion strength to other substrates. Bonding procedures are complex.

Valverde et al. tested plasma treatment on zirconia surfaces. Their results showed an increase of elemental O and a decrease of elemental C [16, 17]. Other authors demonstrated an increase of polarity on zirconia and titanium surfaces after CAP application [18].

8.2.3 Plasma Cleaning and Antimicrobial Effect

Conventional methods of cleaning surfaces include the use of solvents or aggressive chemicals. Plasma technology can selectively produce active species without any toxic residues. As a result, plasma technology could be an interesting alternative for disinfection and decontamination of heat- and moisture-sensitive instruments or surfaces [19–22].

Adherence of *Candida albicans* to acrylic denture base polymers seems to be reduced after treatment by CAP [23]. This could be the result of a direct antimicrobial effect or the modification of prosthetic surface wettability [24].

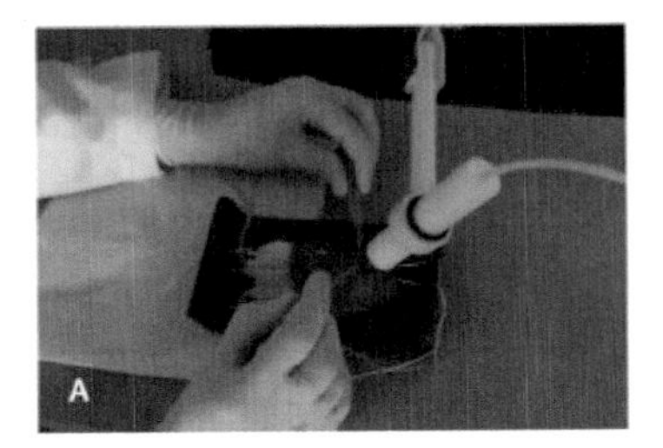

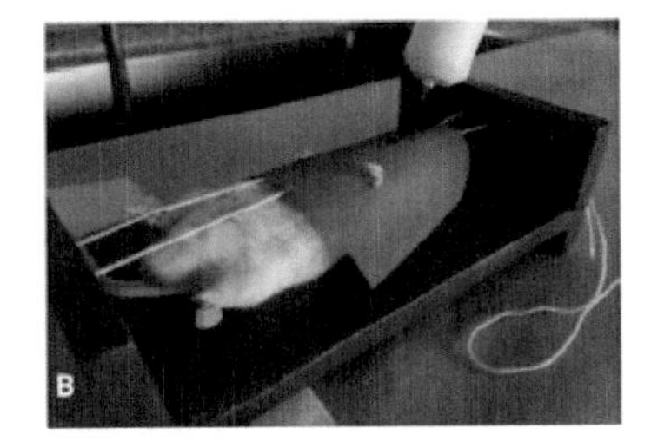

Figure 8.3 Example of an experimental arrangement for in vivo tests [25].

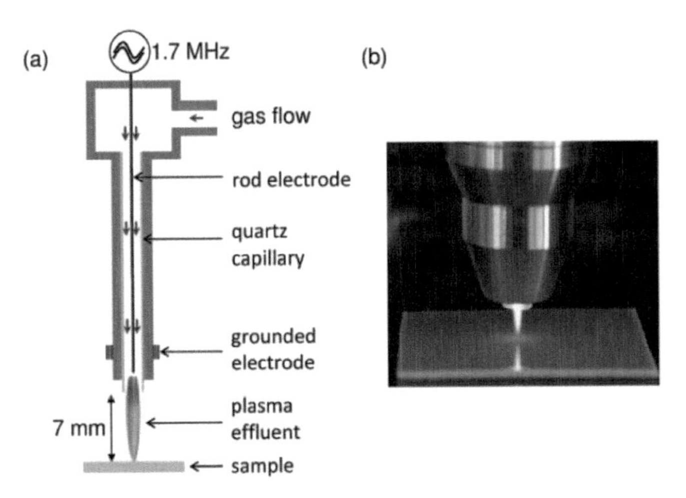

Figure 8.4 Schematic setup and photography of the atmospheric pressure plasma jet (kINPen08, INP Greifswald, Germany) [26].

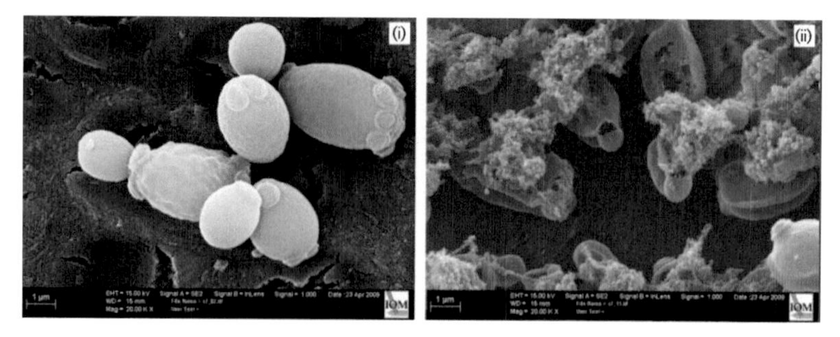

Figure 8.5 Representative scanning electron micrographs of adherent microorganisms on dentin slices: (i) controls without plasma treatment and (ii) after plasma treatment (0.9 s·mm^{-2}). *C. albicans*. magnification ×20,000 [27].

8.3 Direct Application

8.3.1 Increasing the Adhesion

Modern dentistry is based on conservative and minimally invasive techniques. Adhesive restoration is a daily challenge. Bonding of restorations is based on efficient treatment of dental surfaces and careful consideration of bonding protocols. The structure of the enamel allows mechanical microkeying and, thus, reliable and reproducible bonding. The dentin composition is very heterogeneous, making bonding more difficult, the objective being the formation of a hybrid layer between the hydrophilic dentin and the hydrophobic resin.

Direct application of CAP can be achieved on enamel or dentine to enhance bonding. The increased wettability of the dentin surface seems to play a key role [28]. Treatment time influences the interfacial bonding strength (Fig. 8.6) [29].

Recently, Imiolczyk et al. studied the effect of CAP treatment on the strength of adhesive/dentin interfacial bonding. They concluded an improvement in the bonding strength but indicated it depends on the degree of the dentine's mineralization. The CAP sources (plasma jet/dielectric barrier discharge [DBD]) also influence the results [30].

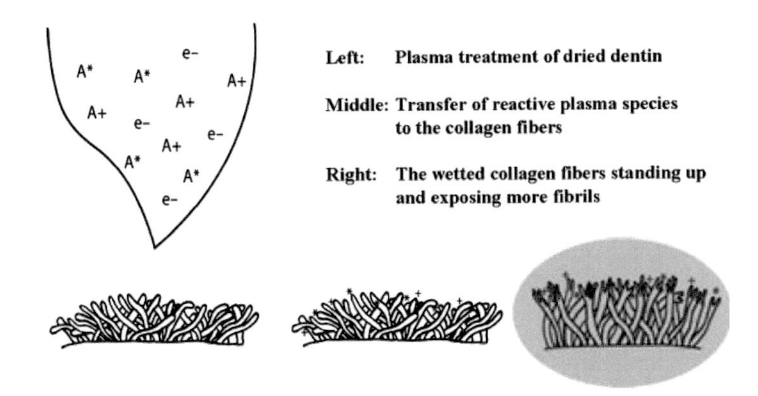

Figure 8.6 Schematic illustration of the effects of plasma treatment on a demineralized dentin surface [29].

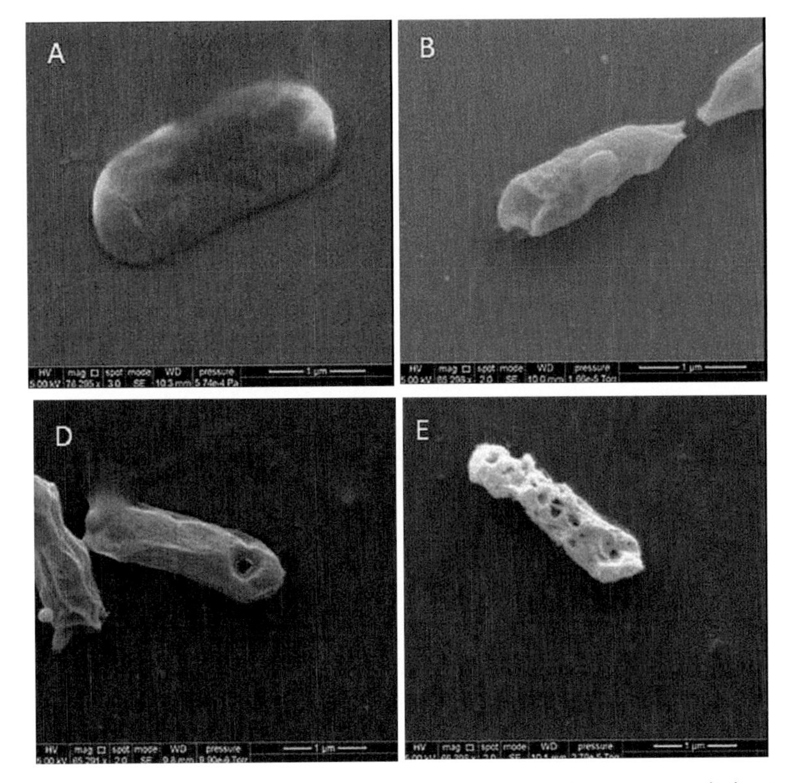

Figure 8.7 Effect of nitrogen afterglow exposure on bacteria morphology (*E. coli*) [31].

8.3.2 Antimicrobial Effects

Antimicrobial effects of CAP are of interest because of the important field of infectious pathology in odontology. Plasma removal of infected dentine is of interest because of the lack of heat damage to the dental pulp [27, 32]. Oral pathogens (biofilm or planktonic state) can be killed by the application of CAPs [25, 26, 33–35]. Reduction of oral microorganisms' adhesion to dentine has been demonstrated in vitro. This is relevant to the treatment of caries and periodontal disease. Decontamination of implant surfaces and nonsurgical treatment of peri-implantitis are challenging. CAP could be a new alternative [36].

Exact mechanisms of plasma antimicrobial effects are not clearly elucidated today. One hypothesis is the generation of reactive species, such as oxygen, nitrogen, and nitrogen oxide radicals. Another hypothesis is the etching effect of the plasma on biofilms.

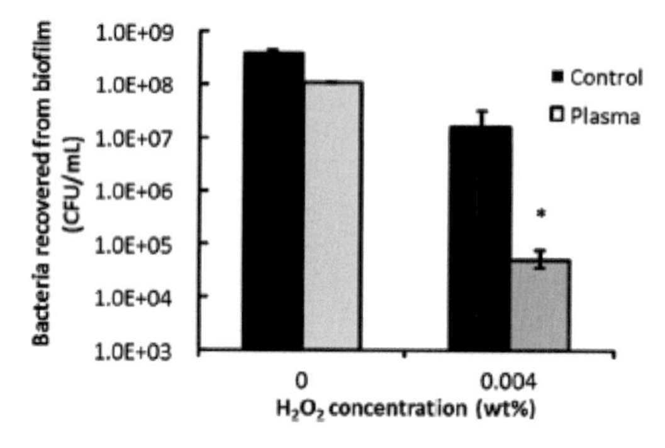

Figure 8.8 Biofilm responses to oxidative stresses [32].

8.3.3 Endodontic Applications of Cold Atmospheric Plasmas

8.3.3.1 Plasma and disinfection

Root canal bacteria cause periapical bone lesions. The aim of endodontic treatment is to prevent and to cure apical periodontitis. To attain this goal, endodontic treatment is based on the exclusion of microorganisms from the root canal system. Disinfection is complicated by the nature of root canal microorganisms organized in the biofilm (*Enterococcus faecalis*, *Fusobacterium nucleatum*, etc.) and due to the fact that they may be lodged in areas inaccessible to instruments (dentinal tubules, isthmus, lateral canal, etc.). Conventional endodontic procedures to eliminate microorganisms are mechanical shaping, hypochlorite irrigation, and laser irradiation. CAP is an effective therapy in endodontics for its direct effects on biofilm. Also, CAP seems to increase wettability of the dentine wall, which improves the effectiveness of usual antimicrobial irrigation. However, eliminating an endodontic biofilm is still challenging [37–40].

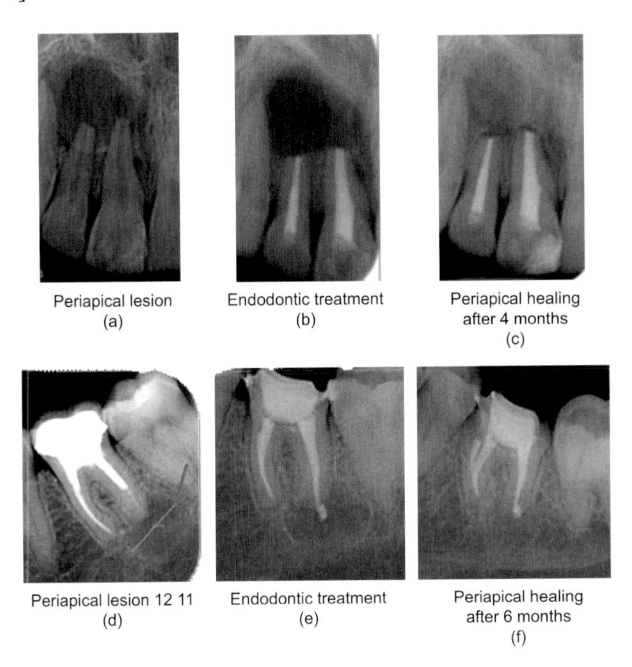

Figure 8.9 (a–c) Primary endodontic treatment. (d–f) Secondary endodontic treatment. All images courtesy of Dr. Marie Georgelin-Gurgel.

8.3.3.2 Plasma and dentine wall

CAP is safe for the dentine wall. Its low temperature does not affect the microhardness and roughness of the root canal dentin.

Long-term success of endodontic treatment is tightly connected with the quality of the canal and coronal filling. An endodontic sealer and condensed gutta-percha efficiently seal the root canal system. Coronal leakage is currently mentioned in endodontic treatment failure. CAP increases the mechanical properties of the adhesive-dentin interface, facilitating adhesive permeation into dentinal tubules. Thus, CAP used before coronal restoration can contribute to endodontic long-term success [40–42].

8.3.4 Tooth Bleaching

Tooth bleaching is an esthetic service in odontology. Hydrogen peroxide (30%–44% H_2O_2) is used in this indication. Usually, bleaching is enhanced by eating H_2O_2, which could be less safe for the tooth. CAPs could be complementary or alternative to the conventional method. This new technique may lead to thermal damage [43, 44].

8.3.5 Periodontal Treatment

Periodontitis is a chronic multifactorial inflammatory disease resulting from an imbalance between a dysbiotic biofilm and the host's immune response, which can lead to progressive loss of tooth supporting tissues and finally tooth loss [45].

Periodontal treatment usually associates chemical (antiseptic, antibiotic) decontamination of the periodontal lesions with mechanical (ultrasonic, sonic, or laser) biofilm disruption. CAP could be an interesting complement or alternative to conventional periodontal treatment through its bactericidal effect, especially on anerobic bacteria such as *Porphyromonas gingivalis* (one of the major periopathogens) [34].

It seems to have a superior antiplaque effect compared to sonic debridement and no adverse side effects on periodontal ligament stem cells. It could even trigger them into osteoblast differentiation [46]. However, there is still a lack of studies.

In those cases application of CAP could be directly through a plasma jet device or indirectly, by means of plasma-activated water (PAW) [47].

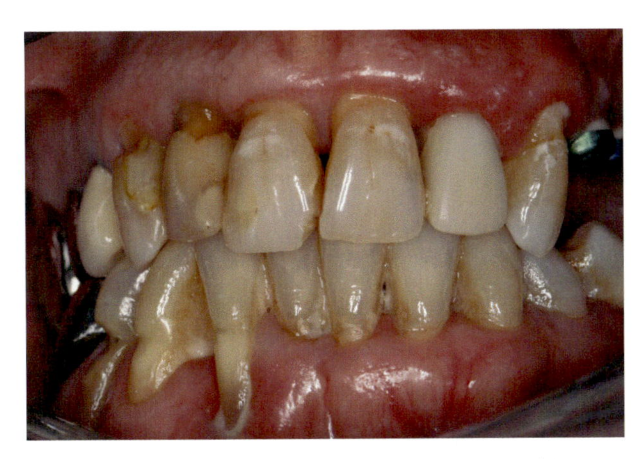

Figure 8.10 Periodontitis, a clinical view. Image courtesy of Sara Laurencin-Dalicieux.

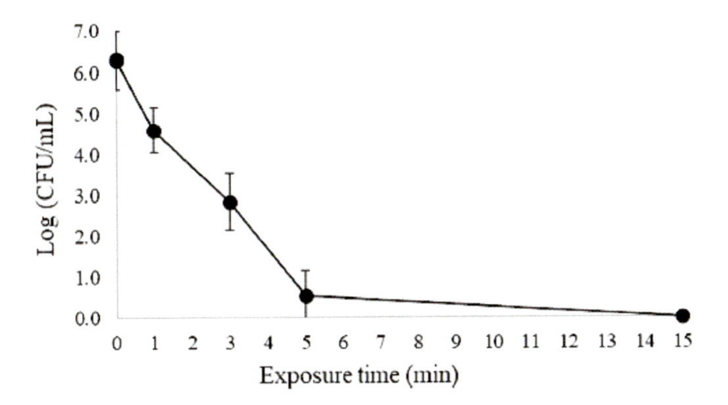

Figure 8.11 Bacterial survival curve. *Pseudomonas aeruginosa* biofilms grown in continuous culture on stainless-steel coupons in AB synthetic medium were treated with DNase and exposed to plasma generated in moistened air [22].

8.3.6 Peri-implantitis Treatment

Peri-implantitis is a plaque-associated pathologic condition occurring in the tissue around dental implants and is characterized by inflammation in the peri-implant mucosa and subsequent progressive loss of supporting bone. Peri-implant mucositis, an inflammation of the peri-implant mucosa, usually precedes peri-implantitis [48].

The difficulty in peri-implant disease management resides in the fact that on the one hand the biofilm needs to be removed from the implant surface and the peri-implant lesion and on the other hand reosteogration of the implant is necessary in order to obtain perfect wound healing. CAP seems promising in this field because it is able to detoxify the implant surface without altering it, unlike more conventional treatments. It also increases implant hydrophilicity and roughness important for cell colonization and seems to have an inductive effect on osteoblasts as well as soft and hard tissue decontamination (Fig. 8.12) [49, 50].

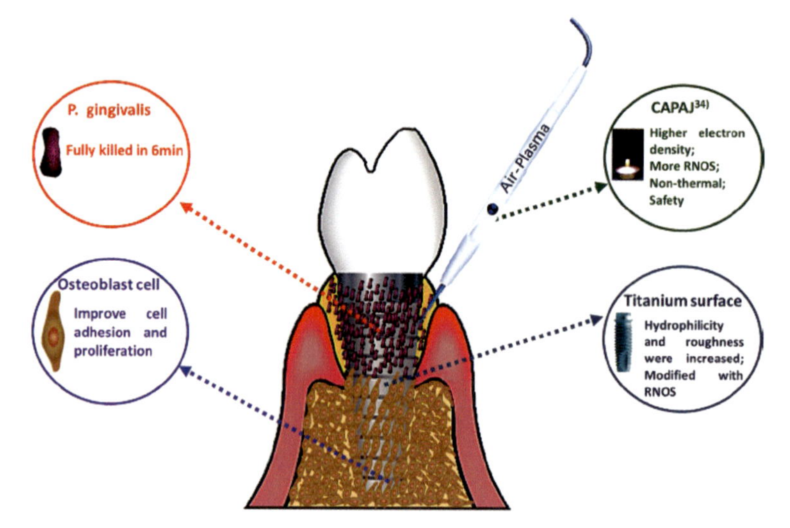

Figure 8.12 Schematic illustration of peri-implantitis treatment by plasma [49].

8.4 Discussion and Conclusion

The majority of oral diseases are infectious ones. The antimicrobial effect of CAP could replace conventional therapeutics or medications. Plasma does not induce bacterial resistance, unlike antibiotics. It also doesn't have toxic residues. Increase of wettability and modification of biologic or prosthetic surfaces is of interest.

It would be interesting to study healing properties of CAPs on the oral mucosa. It would also be interesting to explore plasma drug delivery in dental or periodontal tissues.

References

1. E. Stoffels, A. J. Flikweert, W. W. Stoffels, G. M. W. Kroesen, Plasma needle: a nondestructive atmospheric plasma source for fine surface treatment of (bio) materials, Plasma Sources Sci. Technol., 11 (2002) 383–388.
2. S. Cousty, S. Villeger, J. P. Sarette, A. Ricard, M. Sixou, Inactivation of Escherichia coli in the flowing afterglow of an N2 discharge at reduce pressure: study of the destruction mechanisms of bacteria and hydrodynamics of the afterglow flow, Eur. Phys. J. Appl. Phys., 34 (2006) 143–146.
3. A. Noro, M. Kaneko, I. Murata, M. Yoshinari, Influence of surface topography and surface physicochemistry on wettability of zirconia (tetragonal zirconia polycrystal), J. Biomed. Mater. Res. Part B: Appl. Biomater., 101 (2013) 355–363.
4. W. J. Shon, S. H. Chung, H. K. Kim, G. J. Han, B. H. Cho, Y. S. Park, Peri-implant bone formation of non-thermal atmospheric pressure plasma-treated zirconia implants with different surface roughness in rabbit tibiae, Clin. Oral Implants Res., 25 (2013) 573–579.
5. G. Giro, N. Tovar, L. Witek, C. Marin, N. R. Silva, E. A. Bonfante, et al., Osseointegration assessment of chairside argon-based nonthermal plasma-treated Ca–P coated dental implants, J. Biomed. Mater. Res. Part A, 101 (2013) 98–103.
6. A. Sarkar, D. Pal, S. Sarkar, Cold atmospheric plasma-future of dentistry, IOSR J. Dent. Med. Sci., 17 (2018) 15–20.
7. L. Canullo, T. Genova, N. Naenni, Y. Nakajima, K. Masuda, F. Mussano, Plasma of argon enhances the adhesion of murine osteoblasts on different graft materials, Ann. Anat., 218 (2018) 265–270.
8. H. Naujokat, S. Harder, L. Y. Schulz, J. Wiltfang, C. Florke, Y. Açil, Surface conditioning with cold argon plasma and its effect on the osseointegration of dental implants in miniature pigs, J. Cranio-Maxillo-Facial Surg., 47 (2019) 484e490.
9. D. Ujino, H. Nishizaki, S. Higuchi, S. Komasa, J. Okazaki, Effect of plasma treatment of titanium surface on biocompatibility, Appl. Sci., 9 (2019) 2257.
10. A. Dubuc, P. Monsarrat, F. Virard, N. Merbahi, J. P. Sarrette, S. Laurencin-Dalicieux, S. Cousty, Use of cold-atmospheric plasma in oncology: a concise systematic review, Ther. Adv. Med. Oncol., 10 (2018) 1–12.

11. J. S. Da Silva, S. C. Amico, A. O. Rodrigues, C. A. Barboza, C. Alves Jr., A. T. Croci, Osteoblastlike cell adhesion on titanium surfaces modified by plasma nitriding, Int. J. Oral Maxillofac. Implants, 26 (2011) 237–244.
12. T. Hayakawa, M. Yoshinari, K. Nemoto, Characterization and protein-adsorption behavior of deposited organic thin film onto titanium by plasma polymerization with hexamethyldisiloxane, Biomaterials, 25 (2004) 119–127.
13. U. Walschus, A. Hoene, M. Patrzyk, B. Finke, M. Polak, S. Lucke, et al., Serum profile of pro- and anti-inflammatory cytokines in rats following implantation of low-temperature plasma-modified titanium plates, J. Mater. Sci. Mater. Med., 23 (2012) 1299–1307.
14. K. Schroder, B. Finke, A. Ohl, F. Luthen, C. Bergemann, B. Nebe, et al., Capability of differently charged plasma polymer coatings for control of tissue interactions with titanium surfaces, J. Adhes. Sci. Technol., 24 (2010) 1191–1205.
15. G. J. Han, S. N. Chung, B. H. Chun, C. K. Kim, K. H. Oh, B. H. Cho, Effect of the applied power of atmospheric pressure plasma on the adhesion of composite resin to dental ceramic. J. Adhes. Dent., 14 (2012) 461–469.
16. G. B. Valverde, P. G. Coelho, M. N. Janal, F. C. Lorenzoni, R. M. Carvalho, V. P. Thompson, et al., Surface characterisation and bonding of Y-TZP following non-thermal plasma treatment, J. Dent., 41 (2013) 51–59.
17. P. Güers, S. Wille, T. Strunskus, O. Polonskyi, M. Kern, Durability of resin bonding to zirconia ceramic after contamination and the use of various cleaning methods, Dent. Mater., 35 (2019) 1388–1396.
18. N. R. F. A. Silva, P. G. Coelho, G. B. Valverde, K. Becker, R. Ihrke, A. Quade, et al., Surface characterization of Ti and Y-TZP following non-thermal plasma exposure, J. Biomed. Mater. Res. Part B, 99B (2011) 199–206.
19. L. C. Farrar, D. P. Haack, S. F. McGrath, J. C. Dickens, E. A. O'Hair, J. A. Fralick, Rapid decontamination of large surface areas, IEEE Trans. Plasma Sci., 28 (2000) 173–179.
20. G. B. McCombs, M. L. Darby, New discoveries and directions for medical, dental and dental hygiene research: low temperature atmospheric pressure plasma, Int. J. Dent. Hyg., 8 (2010) 10–15.
21. Y. Jiao, F. R. Tay, L.-n. Niu, J.-h. Chen, Advancing antimicrobial strategies for managing oral biofilm infections, Int. J. Oral Sci., 11 (2019) 28.
22. J. Soler-Arango, C. Figoli, G. Muraca, A. Bosch, G. Brelles-Mariño, The Pseudomonas aeruginosa biofilm matrix and cells are drastically impacted by gas discharge plasma treatment: a comprehensive model

explaining plasma-mediated biofilm eradication, PLoS One, 14(6) (2019) e0216817.

23. M. S. Yildirim, U. Hasanreisoglu, N. Hasirci, N. Sultan, Adherence of Candida albicans to glow-discharge modified acrylic denture base polymers, J. Oral Rehabil., 32 (2005) 518–525.
24. C. A. Zamperini, A. L. Machado, C. E. Vergani, A. C. Pavarina, E. T. Giampaolo, N. C. da Cruz, Adherence in vitro of Candida albicans to plasma treated acrylic resin. Effect of plasma parameters, surface roughness and salivary pellicle, Arch. Oral Biol., 55 (2010) 763–770.
25. A. C. Borges, G. de Morais Gouvêa Lima, T. M. C. Nishime, A. V. L. Gontijo, K. G. Kostov, C. Y. Koga-Ito, Amplitude-modulated cold atmospheric pressure plasma jet for treatment of oral candidiasis: in vivo study, PLoS One, 13(6) (2018) e0199832.
26. K. Fricke, I. Koban, H. Tresp, L. Jablonowski, K. Schroder, A. Kramer, et al., Atmospheric pressure plasma: a high-performance tool for the efficient removal of biofilms, PLoS One, 7 (2012) e42539.
27. S. Rupf, A. Lehmann, M. Hannig, B. Schafer, A. Schubert, U. Feldmann, et al., Killing of adherent oral microbes by a non-thermal atmospheric plasma jet, J. Med. Microbiol., 59 (2010) 206–212.
28. G.-J. Han, J.-H. Kim, S.-N. Chung, B.-H. Chun, C.-K. Kim, D.-G. Seo, H.-H. Son, B.-H. Cho, Effects of non-thermal atmospheric pressure pulsed plasma on the adhesion and durability of resin composite to dentin, Eur. J. Oral Sci., 122 (2014) 417–423.
29. A. C. Ritts, H. Li, Q. Yu, C. Xu, X. Yao, L. Hong, et al., Dentin surface treatment using a non-thermal argon plasma brush for interfacial bonding improvement in composite restoration, Eur. J. Oral Sci., 118 (2010) 510–516.
30. S. M. Imiolczyk, M. Hertel, I. Hase, S. Paris, U. Blunck, S. Hartwig, et al., The influence of cold atmospheric plasma irradiation on the adhesive bond strength in non-demineralized and demineralized human dentin: an study, Open Dent. J., 12 (2018) 960–968.
31. H. Zerrouki, V. Rizzati, C. Bernis, A. Nègre-Salvayre, J. P. Sarrette, S. Cousty. Escherichia coli morphological changes and lipid A removal induced by reduced pressure nitrogen afterglow exposure, PLoS One, 10(4) (2015) e0116083.
32. Q. Honga, X. Donga, M. Chenb, H. Sunc, L. Hongd, Y. Wange, H. Lia, Q. Yua, An in vitro and in vivo study of plasma treatment effects on oral biofilms, J. Oral Microbiol., 11 (2019) 1603524.

33. H. Yamazaki, T. Ohshima, Y. Tsubota, H. Yamaguchi, J. A. Jayawardena, Y. Nishimura, Microbicidal activities of low frequency atmospheric pressure plasma jets on oral pathogens, Dent. Mater. J., 30 (2011) 384–391.

34. Y. Hirano, M. Hayashi, M. Tamura, F. Yoshino, A. Yoshida, M. Masubuchi, K. Imai, B. Ogiso, Singlet oxygen generated by a new non thermal atmospheric pressure air plasma device exerts a bactericidal effect on oral pathogens, J. Oral Sci., 61(4) (2019) 521–525.

35. O. Handorf, T. Weihe, S. Bekeschus, A. C. Graf, U. Schnabel, K. Riedel, J. Ehlbeck, Nonthermal plasma jet treatment negatively affects the viability and structure of Candida albicans SC5314 biofilms, Appl. Environ. Microbiol., 84 (2018) e01163-18.

36. A. N. Idlibi, F. Al-Marrawi, M. Hannig, A. Lehmann, A. Rueppell, A. Schindler, et al., Destruction of oral biofilms formed in situ on machined titanium (Ti) surfaces by cold atmospheric plasma, Biofouling, 29 (2013) 369–379.

37. J. Pan, K. Sun, Y. Liang, P. Sun, X. Yang, J. Wang, et al., Cold plasma therapy of a tooth root canal infected with Enterococcus faecalis biofilms in vitro. J. Endodont., 39 (2013) 105–110.

38. A. Armand, M. Khani, M. Asnaashari, A. Ali Ahmadi, B. Shokri, Comparison study of root canal disinfection by cold plasma jet and photodynamic therapy, Photodiagn. Photodyn. Ther., 26 (2019) 327–333.

39. I. Prada, P. Micó-Muñoz, T. Giner-Lluesma, P. Micó-Martínez, S. Muwaquet-Rodríguez, A. Albero-Monteagudo, Update of the therapeutic planning of irrigation and intracanal medication in root canal treatment. A literature review. J. Clin. Exp. Dent., 11(2) (2019) e185–e193.

40. Y. Li, K. Sun, G. Ye, Y. Liang, H. Pan, G. Wang, Y. Zhao, J. Pan, J. Zhang, J. Fang, Evaluation of cold plasma treatment and safety in disinfecting 3-week root canal Enterococcus faecalis biofilm in vitro, J. Endod., 41(8) (2015) 1325–1330.

41. S. Tabassum, F. R. Khan, Failure of endodontic treatment: the usual suspects, Eur. J. Dent., 10(1) (2016) 144–147.

42. A. Stancampiano, D. Forgione, E. Simoncelli, R. Laurita, R. Tonini, M. Gherardi, V. Colombo, The effect of cold atmospheric plasma (CAP) treatment at the adhesive-root dentin interface, J. Adhes. Dent., 21(3) (2019) 229–237.

43. J. K. Park, S. H. Nam, H. C. Kwon, A. A. Mohamed, J. K. Lee, G. C. Kim, Feasibility of nonthermal atmospheric pressure plasma for intracoronal bleaching. Int. Endodont. J., 44 (2011) 170–175.
44. B. Çelik, İ. D. Çapar, F. İbiş, N. Erdilek, U. K. Ercan, Deionized water can substitute common bleaching agents for nonvital tooth bleaching when treated with non-thermal atmospheric plasma, J. Oral Sci., 61(1) (2019) 103–110.
45. P. N. Papapanou, M. Sanz, et al., Periodontitis: consensus report of workgroup 2 of the 2017 world workshop on the classification of periodontal and peri-implant diseases and conditions, J. Periodontol., 89(Suppl 1) (2018) S173–S182.
46. B. Kleineidam, M. Nokhbehsaim, J. Deschner, et al., Effect of cold plasma on periodontal wound healing: an in vitro study, Clin. Oral Invest., 23 (2019) 1941–1950.
47. Y. Li, J. Pan, G. Ye, et al., In vitro studies of the antimicrobial effect of non-thermal plasma-activated water as a novel mouthwash, Eur. J. Oral Sci., 125(6) (2017) 463–470.
48. J. Caton, G. Armitage, T. Berglundh, et al., A new classification scheme for periodontal and peri-implant diseases and conditions: introduction and key changes from the 1999 classification, J. Periodontol., 89(Suppl 1) 2018 S1–S8.
49. Y. Yang, J. Guo, X. Zhou, Z. Liu, C. Wang, K. Wang, J. Zhang, Z. Wang, A novel cold atmospheric pressure air plasma jet for peri-implantitis treatment: an in vitro study, Dent. Mater. J., 37(1) (2018) 157–166.
50. S. Preissner, A. C. Poehlmann, A. Schubert, A. Lehmann, T. Arnold, O. Nell, S. Rupf, Ex vivo study comparing three cold atmospheric plasma (CAP) sources for biofilm removal on microstructured titanium, Plasma Med., 9(1) (2019) 1–13.

Index